Advances in Industrial Control

Springer
London
Berlin
Heidelberg
New York
Barcelona
Budapest
Hong Kong
Milan
Paris
Santa Clara
Singapore
Tokyo

Other titles published in this Series:

Intelligent Seam Tracking for Robotic Welding
Nitin Nayak and Asok Ray

Identification of Multivariable Industrial Process for Simulation, Diagnosis and Control
Yucai Zhu and Ton Backx

Nonlinear Process Control: Applications of Generic Model Control
Edited by Peter L. Lee

Microcomputer-Based Adaptive Control Applied to Thyristor-Driven D-C Motors
Ulrich Keuchel and Richard M. Stephan

Expert Aided Control System Design
Colin Tebbutt

Modeling and Advanced Control for Process Industries, Applications to Paper Making Processes
Ming Rao, Qijun Xia and Yiquan Ying

Robust Multivariable Flight Control
Richard J. Adams, James M. Buffington, Andrew G. Sparks and Siva S. Banda

Modelling and Simulation of Power Generation Plants
A.W. Ordys, A.W. Pike, M.A. Johnson, R.M. Katebi and M.J. Grimble

Model Predictive Control in the Process Industry
E.F. Camacho and C. Bordons

H_∞ Aerospace Control Design: A VSTOL Flight Application
R.A. Hyde

Neural Network Engineering in Dynamic Control Systems
Edited by Kenneth Hunt, George Irwin and Kevin Warwick

Neuro-Control and its Applications
Sigeru Omatu, Marzuki Khalid and Rubiyah Yusof

Energy Efficient Train Control
P.G. Howlett and P.J. Pudney

Hierarchical Power Systems Control: Its Value in a Changing Industry
Marija D. Ilic and Shell Liu

System Identification and Robust Control
Steen Tøffner-Clausen

K.F. Man, K.S. Tang, S. Kwong and W.A. Halang

Genetic Algorithms for Control and Signal Processing

With 127 Figures

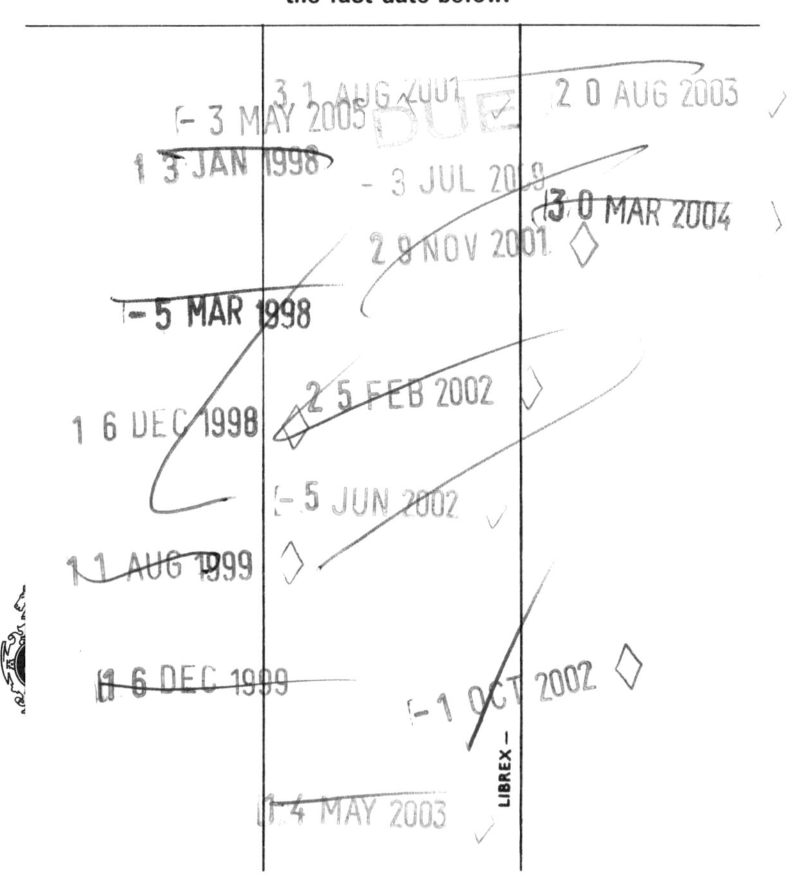

Kim F. Man
Kit S. Tang
Electrical Engineering Department, City University of Hong Kong,
Tat Chee Avenue, Kowloon, Hong Kong

Sam Kwong
Computer Science Department, City University of Hong Kong,
Tat Chee Avenue, Kowloon, Hong Kong

Wolfgang A. Halang
FernUniversitat, Faculty of Electrical Engineering, D-58084 Hagen, Germany

Series Editors

Michael J. Grimble, Professor of Industrial Systems and Director
Michael A. Johnson, Reader in Control Systems and Deputy Director

Industrial Control Centre, Department of Electronic and Electrical Engineering,
Graham Hills Building, 60 George Street, Glasgow G1 1QE, UK

ISBN 3-540-76101-2 Springer-Verlag Berlin Heidelberg New York

British Library Cataloguing in Publication Data
Genetic algorithms for control and signal processing. -
(Advances in industrial control)
1.Genetic algorithms 2.Signal processing 3. Automatic control
I.Man, Kim F.
629.8
ISBN 3540761012

Library of Congress Cataloging-in-Publication Data
A catalog record for this book is available form the Library of Congress

Typesetting: Camera ready by authors
Printed and bound at the Athenæum Press Ltd., Gateshead, Tyne and Wear
69/3830-543210 Printed on acid-free paper

SERIES EDITORS' FOREWORD

The series Advances in Industrial Control aims to report and encourage technology transfer in control engineering. The rapid development of control technology impacts all areas of the control discipline. New theory, new controllers, actuators, sensors, new industrial processes, computer methods, new applications, new philosophies, ..., new challenges. Much of this development work resides in industrial reports, feasibility study papers and the reports of advanced collaborative projects. The series offers an opportunity for researchers to present an extended exposition of such new work in all aspects of industrial control for wider and rapid dissemination.

The emerging technologies in control include fuzzy logic, intelligent control, neural networks and hardware developments like micro-electro-mechanical systems and autonomous vehicles. This volume describes the biological background, basic construction and application of the emerging technology of Genetic Algorithms. Dr Kim Man and his colleagues have written a book which is both a primer introducing the basic concepts and a research text which describes some of the more advanced applications of the genetic algorithmic method. The applications described are especially useful since they indicate the power of the GA method in solving a wide range of problems. These sections are also instructive in showing how the mechanics of the GA solutions are obtained thereby acting as a template for similar types of problems. The volume is a very welcome contribution to the Advances in Industrial Control Series.

M.J. Grimble and M.A. Johnson
Industrial Control Centre
Glasgow, Scotland, UK

PREFACE

The development of Genetic Algorithms (GA) dates back to the early nineteen sixties, but during this period growth in the field was slow. It was not until the exceptional contribution made by Holland in 1975, that an exponential increase of activity in this area first occurred. On entering the nineties we have, suddenly, witnessed another quantum leap in GA research and development by academics and engineers world-wide. The reason for such a phenomenon is easily explained. It has undoubtedly been generated by the advances in solid-state microelectronics fabrication that, in turn, led to the proliferation of widely available, low cost, but speedy computers. The GA as a technique for search and optimization processes, is now ready to be fully exploited and explored.

The GA works on the Darwinian principle of natural selection for which the noted English philosopher, Herbert Spencer, coined the phrase "Survival of the fittest". It is not a mathematically guided problem solver. Therefore, it possesses an intrinsic flexibility and the freedom to choose a desirable solution according to the design specifications. Whether the specifications be nonlinear, constrained, discrete, multimodal, or even NP hard, GA is entirely equal to the challenge. It is, therefore, the aim of this book to gather together relevant GA material that has already been used and demonstrated in various practical applications.

However, considering the wide spectrum of GA applications, we have adopted a strategy of maximized effort to outline the important features of the GA for engineering purposes. We have chosen the areas of control and signal processing to receive specific attention. It is our belief that, the essence of GA characteristics should be clearly outlined, so that the information provided about them is sufficient to enable interested readers to put the knowledge imparted to practical uses.

Within this theme in mind, the book has therefore been designed to be of particular interest to practicing engineers and researchers, although it may also be beneficial to senior-level undergraduates and/or master degree level graduates devoted to the study of GA. It is not our intention to present a series of rigorous mathematical formulae for the GA formulation, nor the genetic concepts that are required to read this book. However, we do recommend that readers acquire the necessary GA background and biological data from other literature, so that they can appreciate the full capability of the GA to solve the types of practical engineering problems that we intend to draw attention to. This assimilation process, of course, must be accompanied by a basic knowledge of computer programming techniques.

The first three chapters of this book are devoted to the mechanism of the GA in search and optimization techniques. Chap. 1 briefly describes the background and biological inspiration of the GA and gives simple examples. Chap. 2 introduces several ways in which to modify the GA formulations for application purposes. The elementary steps necessary to change the genetic operations are presented. The relationship between the objective and fitness functions to determine the quality of the GA evolutionary procedures is discussed. A solid understanding from these two chapters should consolidate the reader's insight into GA utilization and allow him/her to apply this for solving problems in many engineering areas.

In Chap. 3, a number of state-of-the-art techniques are introduced. These procedures are complementary to those described in Chap. 2 but have the ability to solve problems that are considered complex, ill defined and sometime impossible via the use of other gradient types of approach for search and optimization. In this chapter, the parallelism of the GA for the tackling of time consuming computational processes is discussed. Because the GA is not a mathematically oriented scheme, it can therefore, be uniquely applied to solve multiobjective functions in a simultaneous manner. This also applies to system robustness, constrained and multimodal cost functions.

Having formulated the necessary operational procedures and techniques in previous chapters, Chap. 4 involves the application of GA to solve real world problems. There are three cases of study, which are uniquely different in nature. The first case addresses time delay estimation problems. GA is applied to estimate the time delay component which maybe time-variant.

The second study involves Active Noise Control. As the system performance requires the meeting of a speedy response and optimality of quiet zone, the GA parallelism and its multiobjective capability are realized. A final dependable stand-alone microprocessor based architecture that includes the use of FPGA chips is proposed. This hardware framework is not limited to this particular application, but it can be easily converted to other uses if necessary.

The last case in this chapter involves speech recognition systems. GA is used to solve the complex time warping problem. The advantage of GA in this application is its ability to find k-best paths for the speech recognition instead of the precious optimality. Both computational speed and accuracy of the system based on the intrinsic properties of GA have also been carried out.

The second part of this book presents a hierarchical GA structure and its use in solving engineering problems. This is an inspiration finding that arises from the study of biological processes that have already been covered in Chap. 1. A reprise of biological genetics is presented in Chap. 5. The biological relationship between the combination of regulatory sequences and structural genes, as well as the active and inactive genes has been conceptually drawn up. This insight into the genes' structure provides the foundation for the development of hierarchical genetic algorithms (HGA).

In this chapter, a multilayer genetic structure is proposed. The higher level genes controls the genes at the lower level. This process possesses a special feature, in that the genes of the chromosomes are constructed in a hierarchical fashion, while the usual GA operations remain unchanged. In this way, the GA operation remains intact, but its application in solving complex structure optimality is greatly enhanced. This concept is further facilitated by the re-organization of the multiobjective schemes for practical uses.

To reinforce the proposed HGA methodology, both Chaps. 6 and 7 document reports of practical problems that have been tackled by this method. In Chap. 6, a filtering optimization problem in both digital filter design and control systems problem is addressed. In the case of digital filter design, the IIR filter design, regardless of its system performance requirement, can be individually designed and a minimal filter order is reached by the use of HGA. This is then further advanced to tackle the H-infinity control problem. HGA is applied to optimize the weighting functions for the Loop Shaping Design Procedure.

Chap. 7 is an extension to Chap. 6, but here HGA is used to optimize the neural network (NN) structure. This approach is demonstrated by two individual studies. One applies HGA to reach an optimal structure for a number of logic functions. This idea is then extended to address the optimization of neural network topology for a classification system for the diagnosis of breast cancer. In both cases, HGA has been successfully applied.

The second of part of this chapter is the report of a design for fuzzy logic control using the same methodology. In this application, HGA is capable of reaching optimally reduced fuzzy rules and membership functions for control, and yet, the prescribed system performance is maintained. This finding proved that this approach of fuzzy control is very useful particularly for the application of low cost automation.

All the works reported here have been completed by the authors. It is our belief that the fundamental GA material has been aptly demonstrated. We have given enough insight into practical examples for ironing out potential pitfalls in this area. Further work should be continued in the application of HGA concept. Its potential uses in optimizing the topology of computer and telecommunications networks could prove another interesting avenue for research.

K F Man, K S Tang, S Kwong and
W A Halang
July 1996

CONTENTS

LIST OF FIGURES

LIST OF TABLES

CHAPTER 1
INTRODUCTION BACKGROUND AND BIOLOGICAL INSPIRATION

1.1 Biological Background

The fundamental unit of information in living systems is the gene. In general, a gene is defined as a portion of a chromosome that determines or affects a single character or phenotype (visible property), for example, eye colour. It comprises a segment of deoxyribonucleic acid (DNA), commonly packaged into structures called chromosomes. This genetic information is capable of producing a functional biological product which is most often a protein.

1.1.1 Coding of DNA

The basic elements of DNA are nucleotides. Due to their chemical structure, nucleotides can be classified as four different bases, Adenine (A), Guanine (G), Cytosine (C), and Thymine (T). A and G are purines while C and T are pyrimidines. According to Watson and Crick's base pairing theory, G is paired only with C, and A is paired with T (analogue uracil* (U) in ribonucleic acid (RNA)) so that the hydrogen bonds between these pyrimidine-purine pairs are stable and sealed within the complementary strand of DNA organized in form of a double-strand helix [123], see Fig. 1.1.

A triplet code of nucleotide bases specifies the codon, which in turn contains a specific anticodon on transfer RNA (tRNA) and assists subsequent transmission of genetic information in the formation of a specific amino acid. Although there are 64 possible triplet codes, only 20 amino acids are interpreted by codons as tabulated in Table 1.1. It should be noticed that same amino acid may be encoded by different codons in the RNA, and, that there are three codons (UGA, UAA, and UAG) that do not correspond to any amino acid at all, but instead act as signals to stop translation (a process to form polypeptide from RNA).

The organizational hierarchy of DNA can be summarized as in Fig. 1.2.

* T is contained only in DNA and not in RNA. It will be transcribed as another nucleotide, U, in messenger RNA (mRNA).

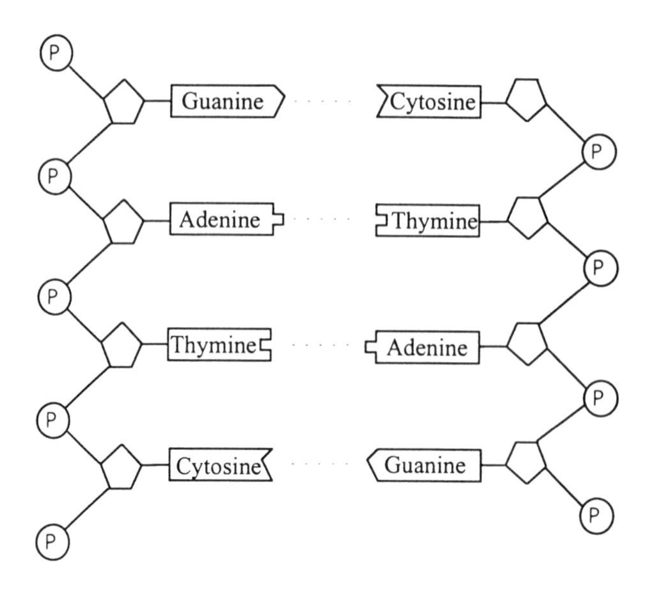

P phosphodiester linkage

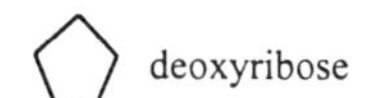

Fig. 1.1. Complementary Structure of Double-Stranded DNA

Table 1.1. The Genetic Code - From Codon to Amino Acid

	Third element in codon			
	U	C	A	G
U U	Phenylalanine	Phenylalanine	Leucine	Leucine
U C	Serine	Serine	Serine	Serine
U A	Phenylalanine	Phenylalanine	stop	stop
U G	Cysteine	Cysteine	stop	Tryptophan
C U	Leucine	Leucine	Leucine	Leucine
C C	Proline	Proline	Proline	Proline
C A	Histidine	Histidine	Glutamine	Glutamine
C G	Arginine	Arginine	Arginine	Arginine
A U	Isolecine	Isolecine	Isolecine	Methionine
A C	Threonine	Threonine	Threonine	Threonine
A A	Asparagine	Asparagine	Lysine	Lysine
A G	Serine	Serine	Arginine	Arginine
G U	Valine	Valine	Valine	Valine
G C	Alanine	Alanine	Alanine	Alanine
G A	Aspartate	Aspartate	Glutamate	Glutamate
G G	Glycine	Glycine	Glycine	Glycine

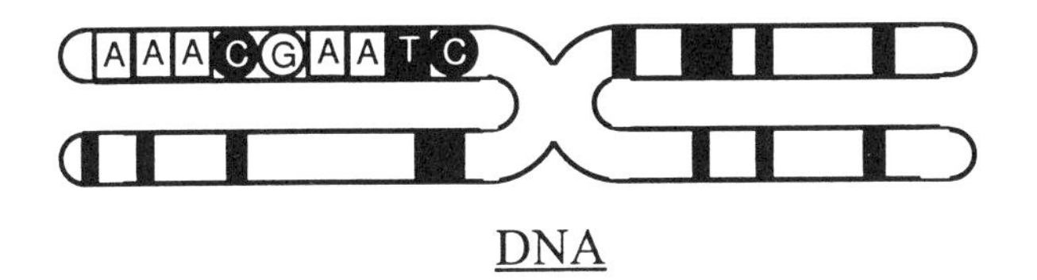

Fig. 1.2. Organizational Hierarchy of DNA

1.1.2 Flow of Genetic Information

There exist three major processes in the cellular utilization of genetic information (Fig. 1.3), replication, transcription and translation.

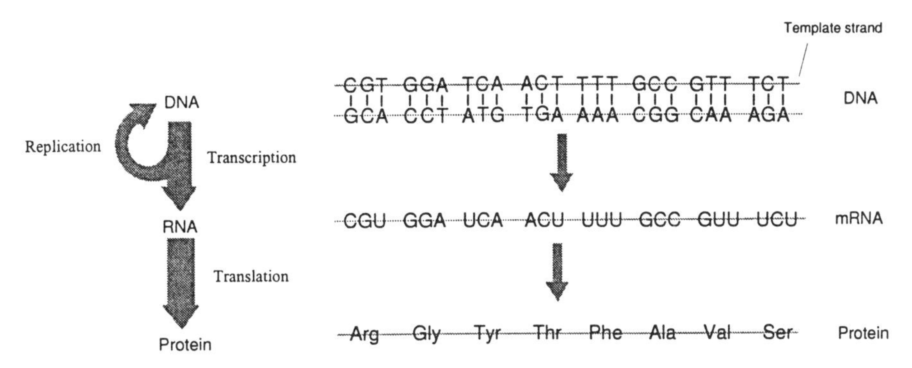

Fig. 1.3. From DNA to Protein

Replication. Genetic information is preserved by DNA replication [83]. During this process, the two parent strands separate, and each serves as a template for the synthesis of a new complementary strand. Each offspring cell inherits one strand of the parental duplex; this pattern of DNA replication is described as semi-conservative.

Transcription. The first step in the communication of genetic information from the nucleus to the cytoplasm is the transcription of DNA into mRNA. During this process, the DNA duplex unwinds and one of the strands serves as a template for the synthesis of a complementary RNA strand mRNA. RNA remains single stranded and functions as the vehicle for translating nucleic acid into protein sequence.

Translation. In the process of translation, the genetic message coded in mRNA is translated on the ribosomes into a protein with a specific sequence of amino acids. Many proteins consist of multiple polypeptide chains. The formulation of polypeptide involves two different types of RNA namely mRNA and tRNA that play important roles in gene translation. Codons are carried by the intermediary formation of mRNA while tRNA, working as an adapter molecule, recognizes codons and inserts amino acids into their appropriate sequential positions in the polypeptide (a product of joining many amino acids). Fig. 1.4 shows the Crick's hypothesis of this translation process.

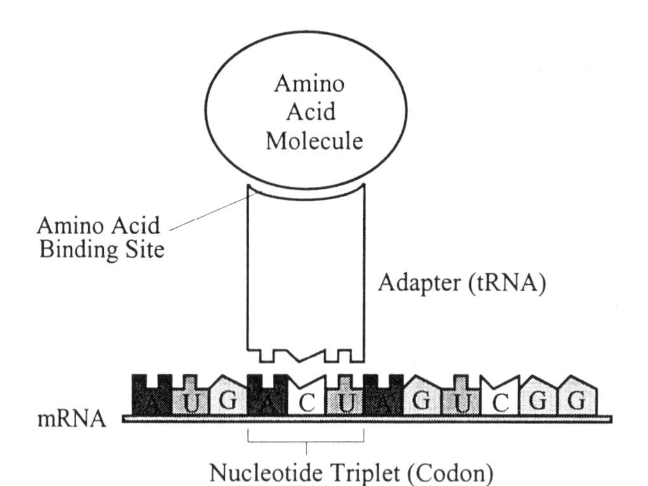

Fig. 1.4. Crick's Hypothesis on tRNA

1.1.3 Recombination

Recombination is a process of the exchange of genetic information. It involves the displacement of a strand in a "host" DNA duplex by a similar strand from a "donor" duplex [116]. Pairing of the displaced host strand with the donor duplex forms a complicated Holliday structure of two duplexes lined by crossed single strands. If the appropriate single strand pairs are broken, the

Holliday structure can be resolved to produce duplexes which have crossed or recombined with one another. The Holliday structure is summarized below:

1. DNA with strand break is aligned with a second homologous DNA, Fig. 1.5a;
2. Reciprocal strand switch produces a Holliday intermediate, Fig. 1.5b;
3. The crossover point moves by branch migration and strand breaks are repaired, Fig. 1.5c;
4. The Holliday intermediate can be cleaved (or resolved) in two ways, producing two possible sets of products. In Fig. 1.5d the orientation of the Holliday intermediate is changed to clarify differences in the two cleavage patterns; and
5. The nonrecombinant and recombinant ends resulting from horizontal and vertical cleavage are shown in Figs. 1.5e and 1.5f, respectively.

Different molecular models of recombinations vary in how they postulate the structure of the host duplex, but all models are based on the ability of the invading strand to pair with its complement [139].

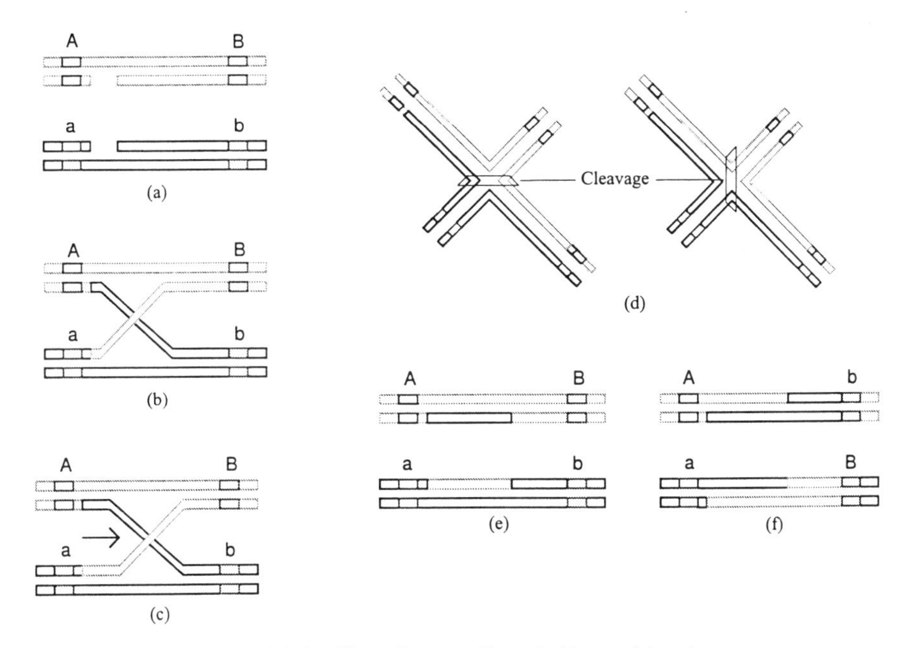

Fig. 1.5. Holliday Model for Homologous Genetic Recombination

1.1.4 Mutation

DNA is a relatively stable polymer and nucleotides generally display a very low tolerance for alterations in genetic information. Very slow, spontaneous relations such as deamination of certain bases, hydrolysis of base-sugar N-glycosidic bonds, formation of pyrimidine dimers (radiation damage), and oxidative damage are critical. An inheritable change in the phenotype; or, from the point of view of molecular biology, any change in a DNA sequence is called a mutation. In general, this operation is rare and random. The process of mutation is blind to its consequences; it throws up every possible combination of mutants, and natural selection then favours those which are better adapted to their environment. Favourable mutations that confer some advantages to the cell in which they occur are rare, being sufficient to provide the variation necessary for natural selection and thus evolution. The majority of mutations, however, are deleterious to the cell.

The most obvious cause of a mutation is an alteration in the coding sequence of the gene concerned. This is often the case with changes which affect only a single base, known as point mutation. Two kinds of point mutation exist: transition mutation and transversion mutation. In a transition mutation (Fig. 1.6b), purines are replaced by purines, and pyrimidines by pyrimidines; i.e. T–A goes to C–G or vice versa. In a transversion mutation (Fig. 1.6c), purines are replaced by purines and pyrimidines; i.e. T–A goes to A–T or G–C, C–G goes to G–C or A–T. Such mutations in coding sequences may be equally well classified by their effects. They may be neutral if there is no effect on coding properties; missence if a codon is changed to another one; or nonsense if the codon changes to a stop codon which means translation will be premature termination. In additions to point mutation, there are frameshift mutations: deletion (Fig. 1.6d), in which one or more base-pairs are lost, and insertion (Fig. 1.6e), in which one or more base-pairs are inserted into the sequence [117].

1.2 Conventional Genetic Algorithm

The basic principles of GA were first proposed by Holland [70]. Thereafter, a series of literature [27, 50, 101] and reports [9, 10, 93, 135, 148, 159] became available. GA is inspired by the mechanism of natural selection where stronger individuals are likely the winners in a competing environment. Here, GA uses a direct analogy of such natural evolution. Through the genetic evolution method, an optimal solution can be found and represented by the final winner of the genetic game.

GA presumes that the potential solution of any problem is an individual and can be represented by a set of parameters. These parameters are regarded as the genes of a chromosome and can be structured by a string of values in binary form. A positive value, generally known as a fitness value, is used to reflect the degree of "goodness" of the chromosome for the problem which

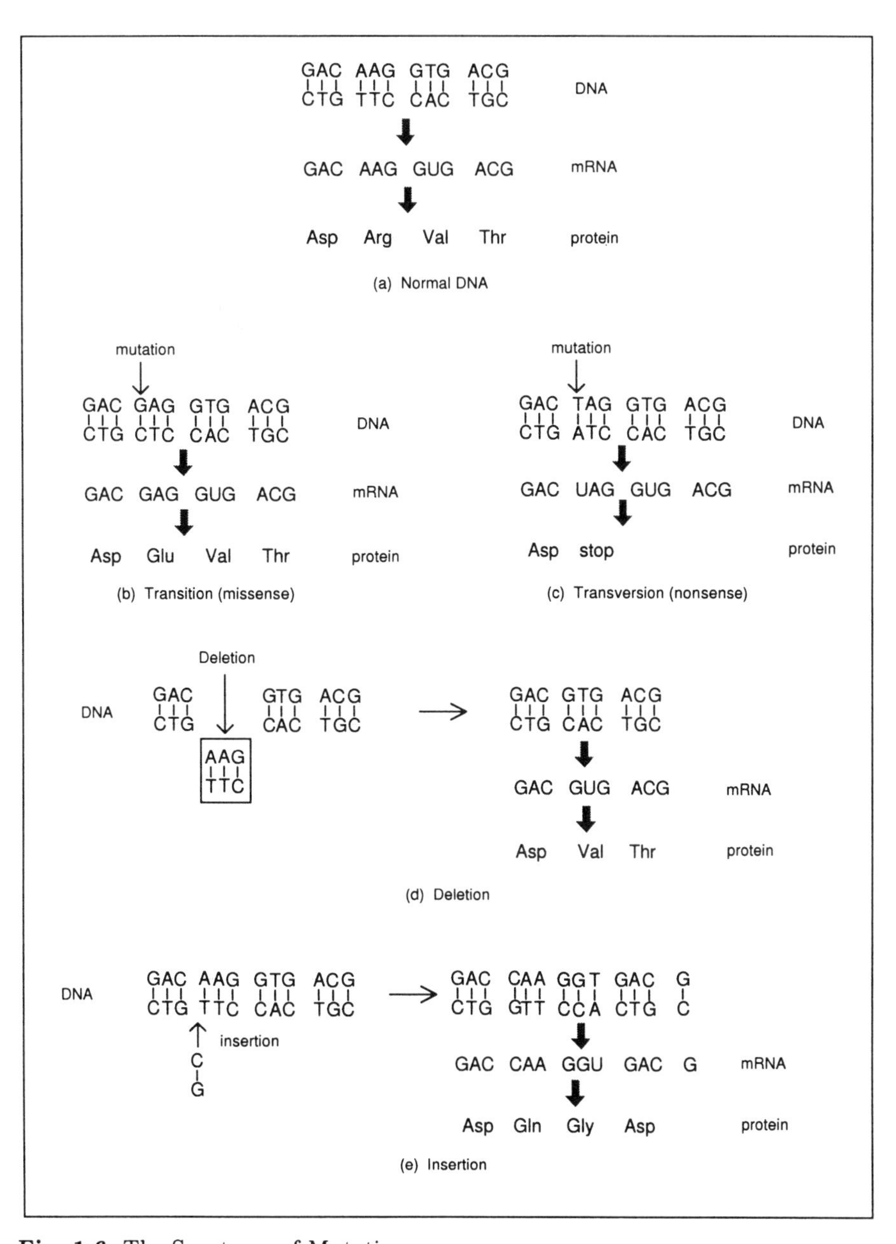

Fig. 1.6. The Spectrum of Mutation

would be highly related with its objective value.

Throughout a genetic evolution, the fitter chromosome has a tendency to yield good quality offspring which means a better solution to any problem. In a practical GA application, a population pool of chromosomes has to be installed and these can be randomly set initially. The size of this population varies from one problem to another although some guidelines are given in [92]. In each cycle of genetic operation, termed as an evolving process, a subsequent generation is created from the chromosomes in the current population. This can only succeed if a group of these chromosomes, generally called "parents" or a collection term "mating pool" is selected via a specific selection routine. The genes of the parents are mixed and recombined for the production of offspring in the next generation. It is expected that from this process of evolution (manipulation of genes), the "better" chromosome will create a larger number of offspring, and thus has a higher chance of surviving in the subsequent generation, emulating the survival-of-the-fittest mechanism in nature.

A scheme called Roulette Wheel Selection [27] is one of the most common techniques being used for such a proportionate selection mechanism. To illustrate this further, the selection procedure is listed in Table 1.2.

Table 1.2. Roulette Wheel Parent Selection

- Sum the fitness of all the population members; named as total fitness (F_{sum}).
- Generate a random number (n) between 0 and total fitness F_{sum}.
- Return the first population member whose fitness, added to the fitness of the preceding population members, is greater than or equal to n.

For example, in Fig. 1.7, the circumference of the Roulette wheel is F_{sum} for all five chromosomes. Chromosome 4 is the fittest chromosome and occupies the largest interval. Whereas chromosome 1 is the least fit which corresponds to a smaller interval within the Roulette wheel. To select a chromosome, a random number is generated in the interval $[0, F_{sum}]$ and the individual whose segment spans the random number is selected.

The cycle of evolution is repeated until a desired termination criterion is reached. This criterion can also be set by the number of evolution cycles (computational runs), or the amount of variation of individuals between

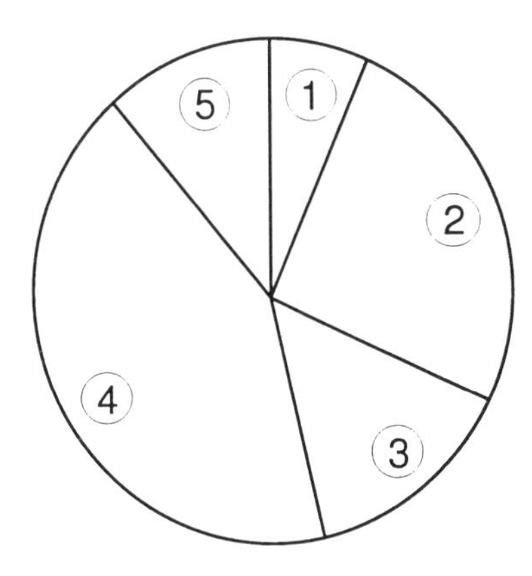

Fig. 1.7. Roulette Wheel Selection

different generations, or a pre-defined value of fitness.

In order to facilitate the GA evolution cycle, two fundamental operators: Crossover and Mutation are required, although the selection routine can be termed as the other operator. To further illustrate the operational procedure, an one-point Crossover mechanism is depicted on Fig. 1.8. A crossover point is randomly set. The portions of the two chromosomes beyond this cut-off point to the right are to be exchanged to form the offspring. An operation rate (p_c) with a typical value between 0.6 and 1.0 is normally used as the probability of crossover.

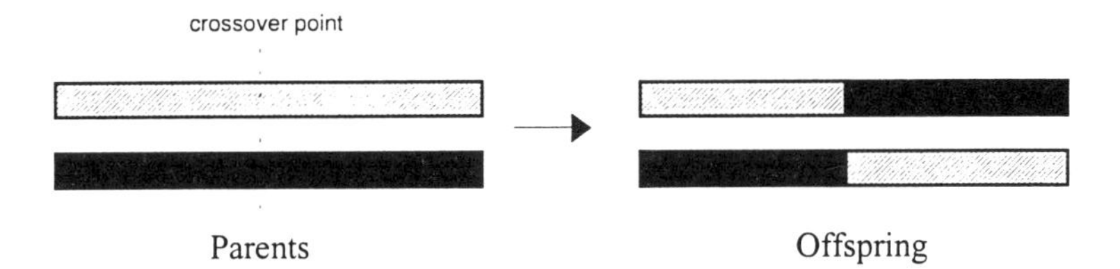

Fig. 1.8. Example of One-Point Crossover

However, for mutation (Fig. 1.9), this applied to each offspring individually after the crossover exercise. It alters each bit randomly with a small probability (p_m) with a typical value of less than 0.1.

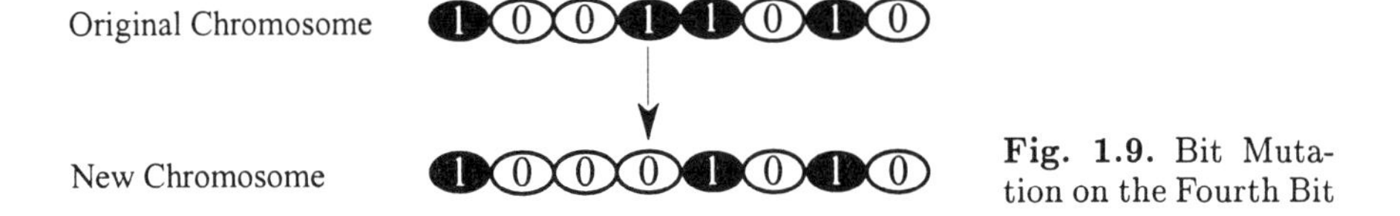

Fig. 1.9. Bit Mutation on the Fourth Bit

The choice of p_m and p_c as the control parameters can be a complex nonlinear optimization problem to solve. Furthermore, their settings are critically dependent upon the nature of the objective function. This selection issue still remains open to suggestion although some guidelines have been introduced by [31, 57]:

- For large population size (100)
  crossover rate: 0.6
  mutation rate: 0.001

- For small population size (30)
  crossover rate: 0.9
  mutation rate: 0.01

Figs. 1.10 and 1.11 summarize the conventional GA.

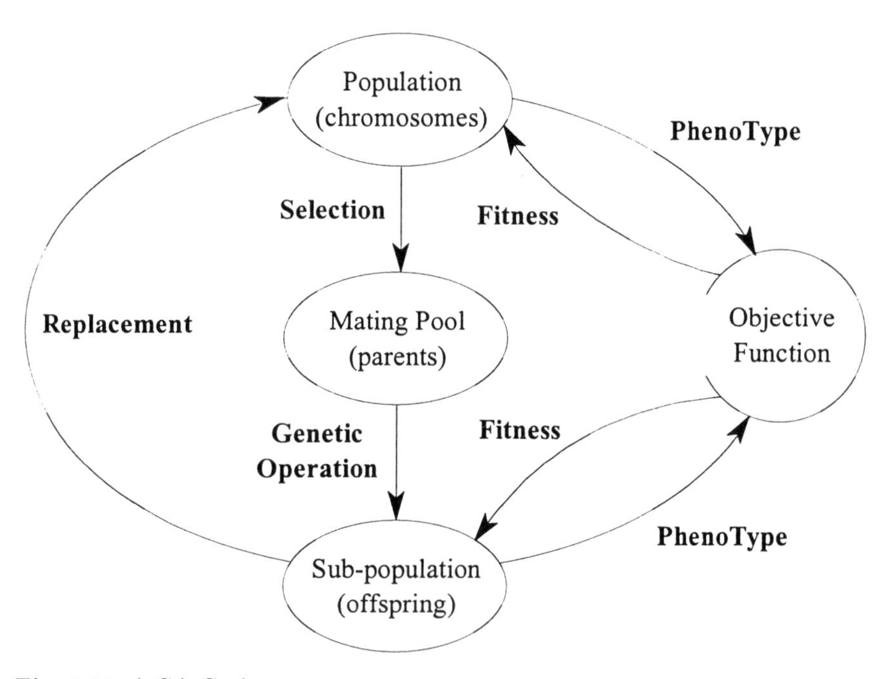

Fig. 1.10. A GA Cycle

```
Genetic Algorithm ()
{
    // start with an initial time
    t := 0;
    // initialize a usually random population of individuals
    init_population P (t);
    // evaluate fitness of all initial individuals of population
    evaluate P (t);
    // evolution cycle
    while not terminated do
            // increase the time counter
            t := t + 1;
            // select a sub-population for offspring production
            P' := select_parents P (t);
            // recombine the "genes" of selected parents
            recombine P' (t);
            // perturb the mated population stochastically
            mutate P' (t);
            // evaluate its new fitness
            evaluate P' (t);
            // select the survivors from actual fitness
            P := survive P,P' (t);
    od
}
```

Fig. 1.11. Conventional Genetic Algorithm Structure

1.3 Theory and Hypothesis

In order to obtain a deeper understanding of GA, it is essential to understand why GA works. At this juncture, there are two schools of thoughts as regards its explanation: Schema Theory and Building Block Hypothesis.

1.3.1 Schema Theory

Consider a simple three-dimensional space, Fig. 1.12, and, assume that the searching space of the solution of a problem can be encoded with three bits, this can be represented as a simple cube with the string 000 at the origin. The corners in this cube are numbered by bit strings and all adjacent corners are labelled by bit strings that differ by exactly 1-bit. If "$*$" represents a "don't care" or "wild card" match symbol, then the front plane of the cube can be represented by the special string $0**$.

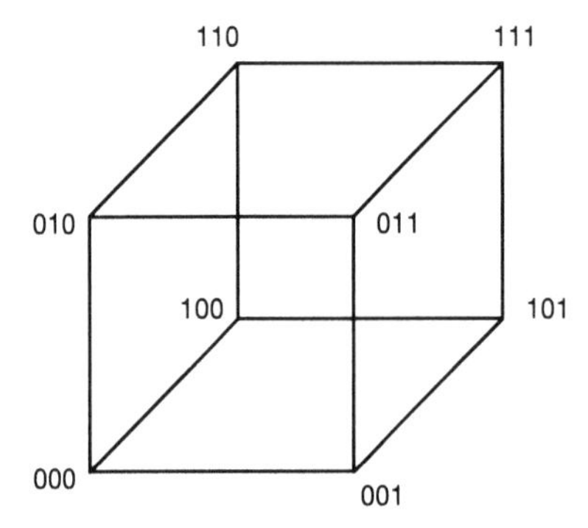

Fig. 1.12. Three-dimensional Cube

Strings that contain "$*$" are referred to as schemata and each schema corresponds to a hyperplane in the search space. A schema represents all strings (a hyperplane or subset of the search space), which match it on all positions other than "$*$". It is clear that every schema matches exactly 2^r strings, where r is the number of *don't care* symbols, "$*$", in the schema template. Every binary encoding is a "chromosome" which corresponds to a corner in the hypercube and is a member of the $2^L - 1$ different hyperplanes, where L is the length of the binary encoding. Moreover, 3^L hyperplanes can be defined over the entire search space.

GA is a population based search. A population of sample points provides information about numerous hyperplanes. Furthermore, low order[†] hyperplanes should be sampled by numerous points in the population. A key part

[†] "order" of a hyperplane refers to the number of actual bit values that appear in the schema.

of the GA's intrinsic or implicit parallelism is derived from the fact that many hyperplanes are sampled when a population of strings is evaluated. Many different hyperplanes are evaluated in an implicitly parallel fashion each time a single string is evaluated, but it is the cumulative effect of evaluating a population of points that provides statistical information about any particular subset of hyperplanes.

Implicit parallelism implies that many hyperplane competitions are simultaneously solved in parallel. The theory suggests that through the process of reproduction and recombination, the schemata of competing hyperplanes increase or decrease their representation in the population according to the relative fitness of the strings that lie in those hyperplane partitions. Because GA operate on populations of strings, one can track the proportional representation of a single schema representing a particular hyperplane in a population. One can also indicate whether that hyperplane will increase or decrease its representation in the population over time, when fitness-based selection is combined with crossover to produce offspring from existing strings in the population. A lower bound on the change in the sampling rate of a single hyperplane from generation t to generation $t+1$ is derived:

Effect of Selection. Since a schema represents a set of strings, we can associate a fitness value $f(S,t)$ with schema "S", and the average fitness of the schema. $f(S,t)$ is then determined by all the matched strings in the population. If *proportional selection* is used in the reproduction phase, we can estimate the number of matched strings of a schema S in the next generation.

Let $\varsigma(S,t)$ be the number of strings matched by schema S at the current generation. The probability of its selection (in a single string selection) is equal to $\frac{f(S,t)}{F(t)}$, where $F(t)$ is the average fitness of the current population. The expected number of occurrences of S in the next generation is

$$\varsigma(S,t+1) = \varsigma(S,t)\frac{f(S,t)}{F(t)} \tag{1.1}$$

Let

$$\varepsilon = \frac{f(S,t) - F(t)}{F(t)} \tag{1.2}$$

If $\varepsilon > 0$, it means that the schema has an above average fitness and vice versa.

Substitute 1.2 into 1.1 and it shows that an "above average" schema receives an exponentially increasing number of strings in the next generations.

$$\varsigma(S,t) = \varsigma(S,0)(1+\varepsilon)^t \tag{1.3}$$

Effect of Crossover. During the evolution of GA, the genetic operations are disruptive to current schemata, therefore, their effects should be considered. Assuming that the length of chromosomes is L and one-point crossover is applied, in general, a crossover point is selected uniformly among $L-1$ possible positions.

This implies that the probability of destruction of a schema S is

$$p_d(S) = \frac{\delta(S)}{L-1} \tag{1.4}$$

or the probability of a schema S survival is

$$p_s(S) = 1 - \frac{\delta(S)}{L-1} \tag{1.5}$$

where δ is the *Defining Length* of the schema S defined as the distance between the outermost fixed positions.

It defines the compactness of information contained in a schema. For example, the *Defining Length* of $*000*$ is 2, while the *Defining Length* of $1*00*$ is 3.

Assuming the operation rate of crossover is p_c, the probability of a schema S survival is:

$$p_s(S) = 1 - p_c \frac{\delta(S)}{L-1} \tag{1.6}$$

Note that a schema S may still survive even if a crossover site is selected between fixed positions, Eqn. 1.6 is modified as

$$p_s(S) \geq 1 - p_c \frac{\delta(S)}{L-1} \tag{1.7}$$

Effect of Mutation. If the bit mutation probability is p_m, then the probability of a single bit survival is $1 - p_m$. Denoting the *Order* of schema S by $o(S)$, the probability of a schema S surviving a mutation (i.e., sequence of one-bit mutations) is

$$p_s(S) = (1 - p_m)^{o(S)} \tag{1.8}$$

Since $p_m << 1$, this probability can be approximated by

$$p_s(S) \approx 1 - o(S)p_m \tag{1.9}$$

Schema Growth Equation. Combining the effect of selection, crossover, and mutation, a new form of the reproductive schema growth equation is derived

$$\zeta(S,t+1) \geq \zeta(S,t)\frac{f(S,t)}{F(t)}\left[1 - p_c\frac{\delta(S)}{L-1} - o(S)p_m\right] \tag{1.10}$$

Based on Eqn. 1.10, it can be concluded that a high average fitness value alone is not sufficient for a high growth rate. Indeed, short, low-order, above-average schemata receive exponentially increasing trials in subsequent generations of a GA.

The *Implicit Parallelism Lower Bound* derived by Holland provides that the number of schemata which are processed in a single cycle is in the order of N^3, where N is the population size. [38] derived the same result and argued that the number of schemata processed was greater than N^3 if $L \geq 64$ and $2^6 \leq N \leq 2^{20}$. This argument does not hold in general for any population size. For a particular string of length L, N must be chosen with respect to L to make the N^3 argument reasonable. In general, the range of values $2^6 \leq N \leq 2^{20}$ does represent a wide range of practical population sizes.

Despite this formulation, it does have its limitations that lead to the restriction of its use. Firstly, the predictions of the GA could be useless or misleading for some problems [61]. Depending on the nature of the objective function, very bad strings can be generated when good building blocks are combined. Such objective functions are referred to as GA-deceptive function [29, 49, 50]. The simplest deceptive function is the minimal deceptive problem, a two-bit function. Assuming that the string "11" represents the optimal solution, the following conditions characterize this problem:

$$\begin{aligned}
f(1,1) &> f(0,0) \\
f(1,1) &> f(0,1) \\
f(1,1) &> f(1,0) \\
f(*,0) &> f(*,1) \quad \text{or} \quad f(0,*) > f(1,*)
\end{aligned} \tag{1.11}$$

The lower order schemata $0*$ or $*0$ does not contain the optimal string 11 as an instance and leads the GA away from 11. The minimal deceptive problem is a partially deceptive function, as both conditions of Eqn. 1.11 are not satisfied simultaneously. In a fully deceptive problem, all low-order schemata containing a suboptimal solution are better than other competing schemata [29]. However, [61] demonstrated that the deceptive problem was not always difficult to solve.

Secondly, the value of $f(S,t)$ in the current population may differ significantly from the value of $f(S,t)$ in the next, since schemata have interfered with each other. Thus, using the average fitness is only relevant to the first population, [58]. After this, the sampling of strings will be biased and the inexactness makes it impossible to predict computational behaviour.

1.3.2 Building Block Hypothesis

A genetic algorithm seeks near-optimal performance through the juxtaposition of short, low-order, high performance schemata, called the building block [101].

The genetic operators, we normally refer to as crossover and mutation, have the ability to generate, promote, and juxtapose (side by side) building blocks to form the optimal strings. Crossover tends to conserve the genetic information present in the strings for crossover. Thus, when the strings for crossover are similar, their capacity to generate new building blocks diminishes. Whereas mutation is not a conservative operator but is capable of generating new building blocks radically.

In addition, parent selection is an important procedure to devise. It tends to be biased towards building blocks that possess higher fitness values, and at the end ensures their representation from generation to generation.

This hypothesis suggests that the problem of coding for a GA is critical to its performance, and that such coding should satisfy the idea of short building blocks.

1.4 A Simple Example

There is no better way to show how a GA works, than to go through a real but simple example to demonstrate its effectiveness.

Problem:

Searching the global maximum point of the following objective function (see Fig. 1.13):

$$z = f(x, y)$$

where $x, y \in [-1, 1]$.

Implementation:

The chromosome was formed by a 16-bit binary string representing x and y co-ordinates each with an eight-bit resolution. One-point crossover and bit mutation were applied with operation rates of 0.85 and 0.1, respectively. The population size was set to four for the purpose of demonstration. (In general,

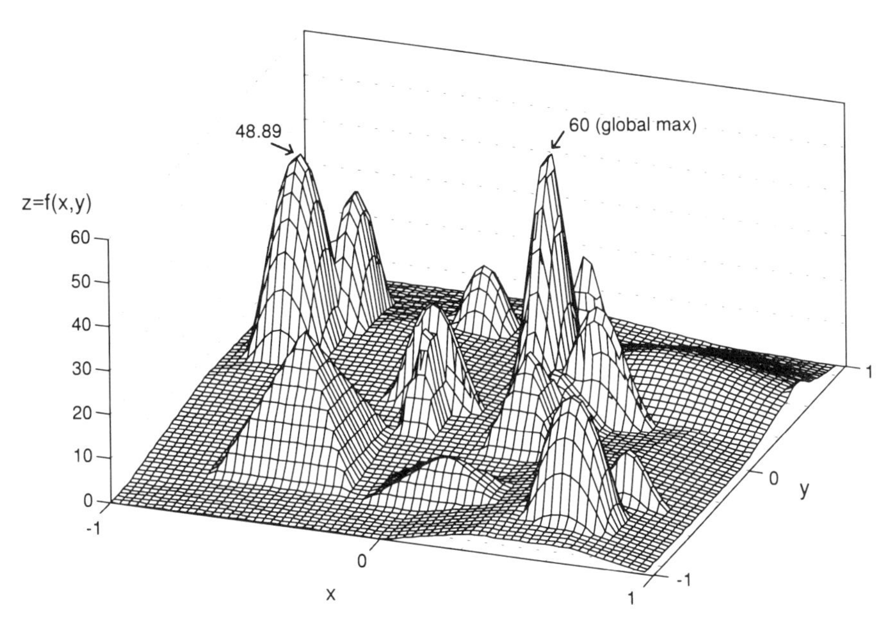

Fig. 1.13. A Multimodal Problem

this number should be much larger). Here, only two offspring were generated for each evolution cycle.

This example was conducted via simulation based on MATLAB with the Genetic Toolbox [18]. Fig. 1.14 shows the typical genetic operations and the changes within the population from first to second generation.

The fitness of the best chromosome during the searching process is depicted on Fig. 1.15 and clearly shows that a GA is capable of escaping from local maxima and finding the global maximum point.

The objective values of the best chromosome in the pool against the generations are depicted in Fig. 1.16.

STEP 1: Parent Selection

	First Population	x	y	Objective Value z = f(x,y)
→	0100110100101000	-0.0740	-0.6233	5.4297
→	0101010110000101	-0.1995	0.9541	0.6696
	0000010100110110	-0.9529	-0.7175	0.2562
	1000101011001011	0.9070	0.1065	4.7937

STEP 2: CROSSOVER

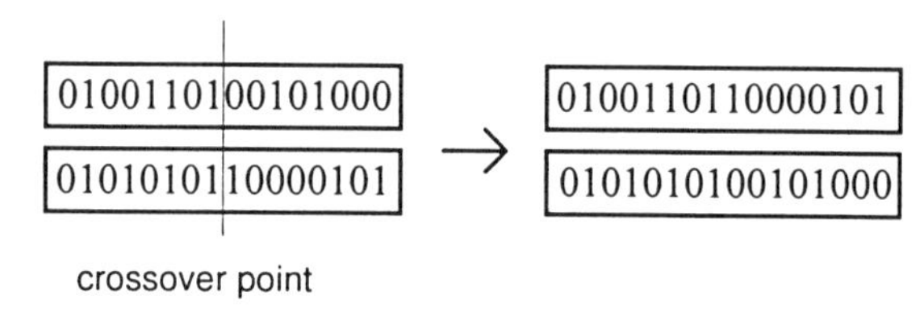

STEP 3: MUTATION

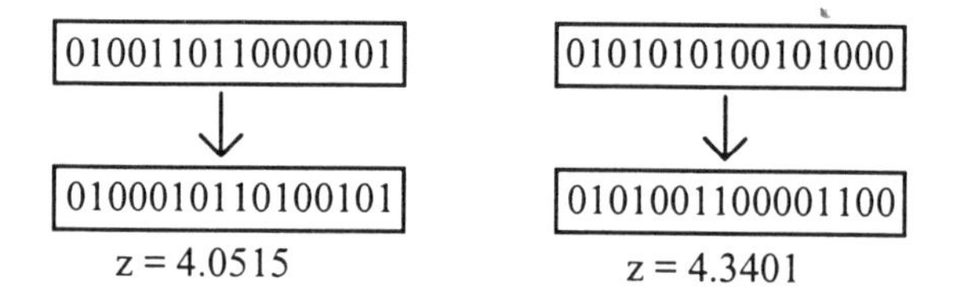

STEP 4: Reinsertion

	First Population	x	y	Objective Value z = f(x,y)
	0100110100101000	-0.0740	-0.6233	5.4297
→	0100010110010101	-0.0504	0.5539	4.0515
→	0101001100001100	-0.2309	-0.9372	4.3401
	1000101011001011	0.9070	0.1065	4.7937

Fig. 1.14. Generation to Generation

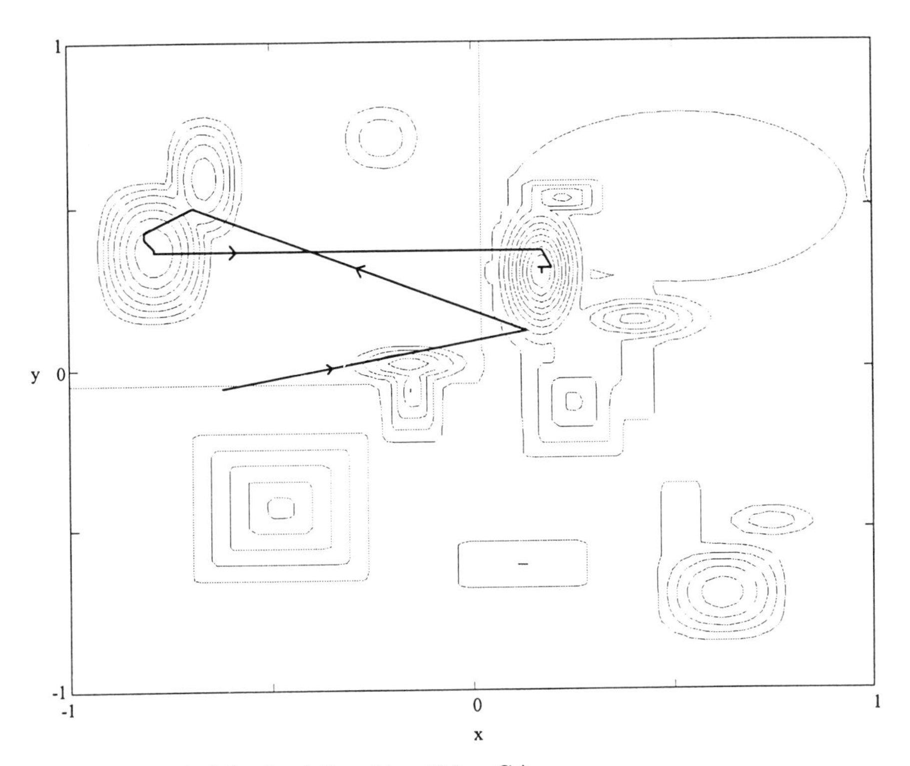

Fig. 1.15. Global Optimal Searching Using GA

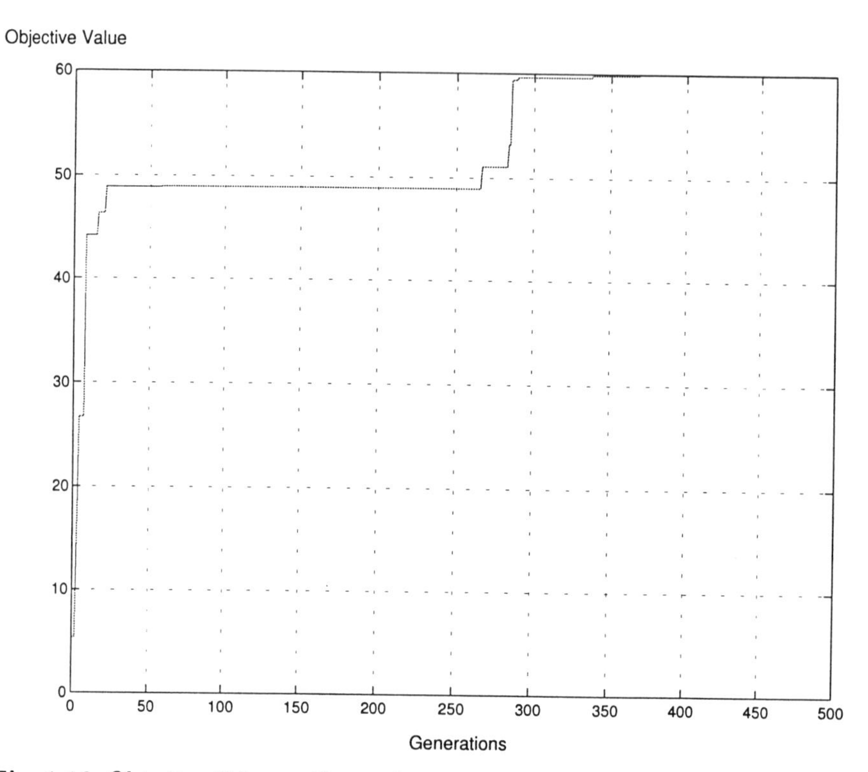

Fig. 1.16. Objective Value vs Generations

CHAPTER 2
MODIFICATION IN GENETIC ALGORITHM

The GA mechanism is neither governed by the use of the differential equations nor does it behave like a continuous function. However, it possesses the unique ability to search and optimize a solution for a complex system, where other mathematical oriented techniques may have failed to compile the necessary design specifications. Due to its evolutionary characteristics, a standard GA may not be flexible enough for a practical application, and an engineering insight is always required whenever a GA is applied. This becomes more apparent where the problem to be tackled is complicated, multi-tasking and conflicting. Therefore, a means of modifying the GA structure is sought in order to meet the design requirements. There are many facets of operational modes that can be introduced. It is the main task of this chapter to outline the essential methodologies.

2.1 Chromosome Representation

The problem to be tackled varies from one to the other. The coding of chromosome representation may vary according to the nature of the problem itself. In general, the bit string encoding [70] is the most classic method used by GA researchers because of its simplicity and traceability. The conventional GA operations and theory (scheme theory) are also developed on the basis of this fundamental structure. Hence, this representation is adopted in many applications. However, one minor modification can be suggested in that a Gary code may be used by the binary coding. [69] investigated the use of GA for optimizing functions of two variables based on a Gray code representation, and discovered that this works slightly better than the normal binary representation.

Recently, a direct manipulation of real-value chromosomes [77, 164] raised considerable interest. This representation was introduced especially to deal with real parameter problems. The work currently taking place [77] indicates that the floating point representation would be faster in computation and more consistent from the basis of run-to-run. At the same time, its performance can be enhanced by special operators to achieve high accuracy [101]. However, the opinion given by [51] suggested that a real-coded

GA would not necessarily yield good result in some situations, despite the fact that many practical problems have been solved by using real-coded GA. So far, there is insufficient consensus to be drawn from this argument.

Another problem-oriented chromosome representation is the order-based representation which is particular useful for those problems where a particular sequence is required to search. The linear linked list is the simplest form for this representation. Normally, each node has a data field and a single link field. The link field points to the address of the successor's data field. This is a chain-like structure which is shown in Fig. 2.1.

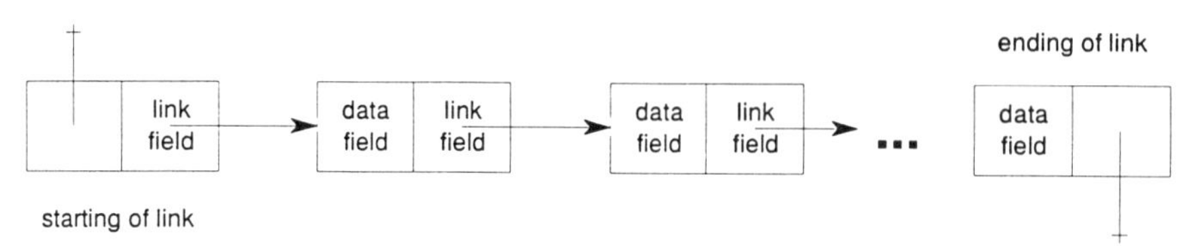

Fig. 2.1. Linear Link List

This type of chromosome formulation can be found in [86] for solving the Dynamic Time Warping system and Robotic Path Planning [101] in which the sequence of the data is essential to represent a solution. In general, the length of the chromosomes in the form of link field may vary. This can be supplemented by a series of special genetic operations in order to meet the design of the GA process. A generalized structure in the form of Graph representation can be introduced. It allows loops to link the other data blocks as indicated in Fig. 2.2. A successful implementation of this structure is demonstrated by solving the graph colouring problem [27].

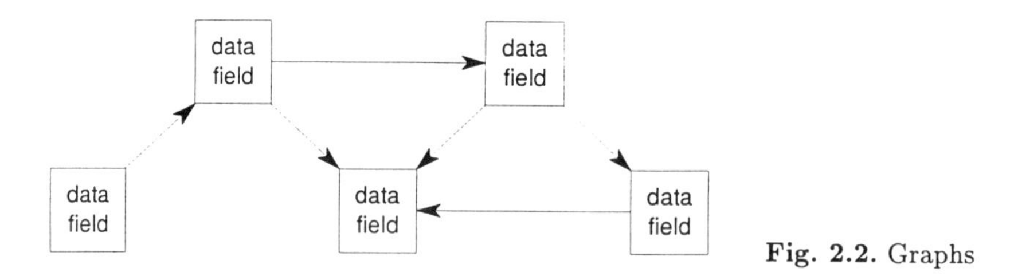

Fig. 2.2. Graphs

These order-based encoding techniques have an important advantage over literal encoding techniques in that they rule out a tremendous number of suboptimal solutions. The process avoids the problem that literal encoding encounters when illegal solutions are often generated by the crossover operations.

In some cases, an index can be used as the chromosome element instead of a real value. A typical example can be given by a look-up table format. This has proved to be a useful technique for nonlinear term selection [39].

All in all, the modification in chromosome representation comprises an endless list and is largely dependent on the nature of individual problems. A well chosen chromosome format can enhance the understanding of the problem formulation and also alleviate the burden of practical implementation.

2.2 Objective and Fitness Functions

An objective function is a measuring mechanism that is used to evaluate the status of a chromosome. This is a very important link to relate the GA and the system concerned. Since each chromosome is individually going through the same evaluating exercise, the range of this value varies from one chromosome to another. To maintain uniformity, the objective value(s) O is mapped into a fitness value(s) [50, 101] with a map Ψ where the domain of F is usually greater than zero.

$$\Psi : O \rightarrow F \tag{2.1}$$

2.2.1 Linear Scaling

The fitness value f_i of chromosome i has a linear relationship with the objective value o_i as

$$f_i = ao_i + b \tag{2.2}$$

where a and b are chosen to enforce the equality of the average objective value and the average fitness value, and cause maximum scaled fitness to be a specified multiple of the average fitness.

This method can reduce the effect of genetic drift by producing an extraordinarily good chromosome. However, it may introduce a negative fitness value which must be avoided in the GA operations [50]. Hence, the choice of a and b are dependent on the knowledge of the range of the objective values.

2.2.2 Sigma Truncation

This method avoids the negative fitness value and incorporates the problem dependent information into the scaling mechanism. The fitness value, f_i of chromosome i is calculated according to

$$f_i = o_i - (\bar{o} - c\sigma) \tag{2.3}$$

where c is a small integer, $\bar{o}$ is the mean of the objective values, σ is the standard deviation in the population.

To prevent negative value of f, any negative result $f < 0$ is arbitrarily set to zero. Chromosomes whose fitness values are less than c (a small integer from the range 1 and 5) standard deviation from the average fitness value are not selected.

2.2.3 Power Law Scaling

The actual fitness value is taken as a specific power of the objective value, o_i

$$f_i = o_i^k \tag{2.4}$$

where k is in general problem dependent or even varying during the run [47].

2.2.4 Ranking

There are other methods that can be used such as the Ranking scheme [7]. The fitness values do not directly relate to their corresponding objective values, but to the ranks of the objective values.

Using this approach can help the avoidance of premature convergence and speed up the search when the population approaches convergence [158]. On the other hand, it requires additional overheads in the GA computation for sorting chromosomes according to their objective values.

2.3 Selection Methods

To generate good offspring, a good parent selection mechanism is necessary. This is a process used for determining the number of trials for one particular individual used in reproduction. The chance of selecting one chromosome as a parent should be directly proportional to the number of offspring produced.

[7] presented three measures of performance of the selection algorithms, *Bias*, *Spread* and *Efficiency*.

- *Bias* defines the absolute difference between individuals in actual and expected probability for selection. Optimal zero bias is achieved when an individual's probability equals its expected number of trials.
- *Spread* is a range in the possible number of trials that an individual may achieve. If $g(i)$ is the actual number of trials due to each individual i, then the "minimum spread" is the smallest spread that theoretically permits zero bias, i.e.

$$g(i) \in \{\lfloor et(i) \rfloor, \lceil et(i) \rceil\} \tag{2.5}$$

where $et(i)$ is the expected number of trials of individual i, $\lfloor et(i) \rfloor$ is the floor and $\lceil et(i) \rceil$ is the ceiling.
Thus, the spread of a selection method measures its consistency.
- *Efficiency* is related to the overall time complexity of the algorithms.

The selection algorithm should thus be achieving a zero bias whilst maintaining a minimum spread and not contributing to an increased time complexity of the GA.

Many selection techniques employ Roulette Wheel Mechanism (see Table 1.2). The basic roulette wheel selection method is a stochastic sampling with replacement (SSR). The segment size and selection probability remain the same throughout the selection phase and the individuals are selected according to the above procedures. SSR tends to give zero bias but potentially inclines to a spread that is unlimited.

Stochastic Sampling with Partial Replacement (SSPR) extends upon SSR by resizing a chromosome's segment if it is selected. Each time an chromosome is selected, the size of its segment is reduced by a certain factor. If the segment size becomes negative, then it is set to zero. This provides an upper bound on the spread of $\lceil et(i) \rceil$ but with a zero lower bound and a higher bias. The roulette wheel selection methods can generally be implemented with a time complexity of the order of $NlogN$ where N is the population size.

Stochastic Universal Sampling (SUS) is another single-phase sampling algorithm with minimum spread, zero bias and the time complexity in the order of N [7]. SUS uses an N equally spaced pointer, where N is the number of selections required. The population is shuffled randomly and a single random number in the range $[0, \frac{F_{sum}}{N}]$ is generated, *ptr*, where F_{sum} is the sum of the individuals' fitness values. The N individuals are then chosen by generating the N pointers spaced by 1, $[ptr, ptr+1, \ldots, ptr+N+1]$, and selecting those individuals whose fitnesses span the positions of the pointers. An individual is thus guaranteed to be selected a minimum of $\lfloor et(i) \rfloor$ times and no more than $\lceil et(i) \rceil$, thus achieving minimum spread. In addition, as individuals are selected entirely on their position in the population, SUS has zero bias.

2.4 Genetic Operations

2.4.1 Crossover

Although the one-point crossover method was inspired by biological processes, it has one major drawback in that certain combinations of schema cannot be combined in some situations [101].

For example, assume that there are two high-performance schemata:

$$\begin{array}{lcllllllll} S_1 & = & 1 & 0 & 1 & * & * & * & * & 1 \\ S_2 & = & * & * & * & * & 1 & 1 & * & * \end{array}$$

There are two chromosomes in the population, I_1 and I_2, matched by S_1 and S_2, respectively:

$$\begin{array}{lcllllllll} I_1 & = & 1 & 0 & 1 & 1 & 0 & 0 & 0 & 1 \\ I_2 & = & 0 & 1 & 1 & 0 & 1 & 1 & 0 & 0 \end{array}$$

If only one-point crossover is performed, it is impossible to obtain the chromosome that can be matched by the following schema (S_3) as the first schema will be destroyed.

$$\begin{array}{lcllllllll} S_3 & = & 1 & 0 & 1 & * & 1 & 1 & * & 1 \end{array}$$

A multi-point crossover can be introduced to overcome this problem. As a result, the performance of generating offspring is greatly improved. One example of this operation is depicted in Fig. 2.3 where multiple crossover points are randomly selected.

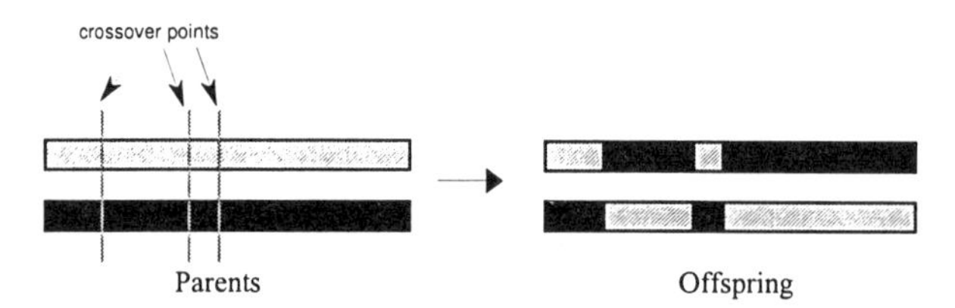

Fig. 2.3. Example of Multi-Point Crossover

Assuming that two-point crossover is performed on I_1 and I_2 as demonstrated below, the resulting offspring are shown as I_3 and I_4 in which I_3 are matched by S_3.

$$\begin{array}{lcllll|ll|ll} I_1 & = & 1 & 0 & 1 & 1 & 0 & 0 & 0 & 1 \\ I_2 & = & 0 & 1 & 1 & 0 & 1 & 1 & 0 & 0 \end{array}$$

$$\begin{array}{lcllllllll} I_3 & = & 1 & 0 & 1 & 1 & 1 & 1 & 0 & 1 \\ I_4 & = & 0 & 1 & 1 & 0 & 0 & 0 & 0 & 0 \end{array}$$

Another approach is the uniform crossover. This generates offspring from the parents, based on a randomly generated crossover mask. The operation is demonstrated in Fig. 2.4.

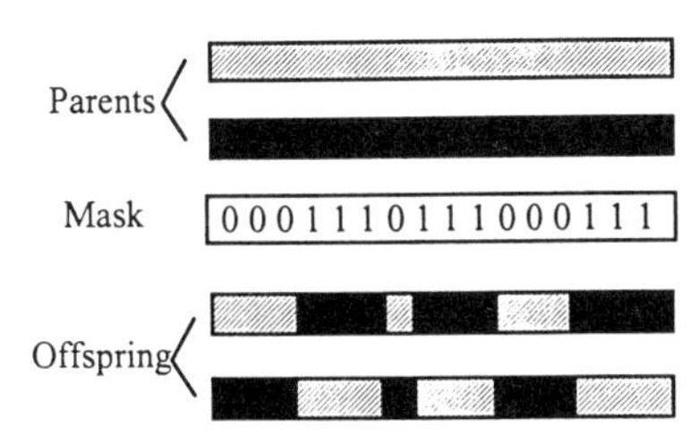

Fig. 2.4. Example of Uniform Crossover

The resultant offspring contain a mixture of genes from each parent. The number of effective crossing points is not fixed, but will be averaged at $L/2$ (where L is the chromosome length).

The preference for using which crossover techniques is still arguable. However, [30] concluded that a two-point crossover seemed to be an optimal number for multi-point crossover. Since then, this has been contradicted by [134] as a two-point crossover could perform poorly if the population has largely being converged because of any reduced crossover productivity. This low crossover productivity problem can be resolved by the utilization of the reduce-surrogate crossover [12].

Since the uniform crossover exchanges bits rather than segments, it can combine features regardless of their relative location. This ability may outweigh the disadvantage of destroying building block solutions and make uniform crossover superior for some problems [137]. [37] reported on several experiments for various crossover operators. A general comment was that each of these crossovers was particularly useful for some classes of problems and quite poor for others, and that the one-point crossover was considered a "loser" experimentally.

Crossover operations can be directly adopted into the chromosome with real number representation. The only difference would be if the string is composed of a series of real numbers instead of binary number.

Some other problem-based crossover techniques have been proposed. [24] designed an "analogous crossover" for the robotic trajectory generation. Therefore, the use of the crossover technique to improve the offspring production, is very much problem oriented. The basic concept in crossover is to exchange gene information between chromosomes. An effective design of crossover operation would greatly increase the convergency rate of a problem.

2.4.2 Mutation

Originally, mutation was designed only for the binary-represented chromosome. To adopt the concept of introducing variations into the chromosome, a random mutation [101] has been designed for the real number chromosome:

$$g = g + \psi(\mu, \sigma) \tag{2.6}$$

where g is the real value gene; ψ is a random function which may be Gaussian or normally distributed; μ, σ are the mean and variance related with the random function, respectively.

2.4.3 Operational Rates Settings

The choice of an optimal probability operation rate for crossover and mutation is another controversial debate for both analytical and empirical investigations. The increase of crossover probability would cause the recombination of building blocks to rise, and at the same time, it also increases the disruption of good chromosomes. On the other hand, should the mutation probability increase, this would transform the genetic search into a random search, but would help to reintroduce the lost genetic material.

As each operator probability may vary through the generations, Davis [25] suggested linear variations in crossover and mutation probability, with a decreasing crossover rate during the run while mutation rate was increased. Syswerda [138] imposed a fixed schedule for both cases but Booker [12] utilized a dynamically variable crossover rate which was dependent upon the spread of fitness. [26, 27] modified the operator probabilities according to the success of generating good offspring. Despite all these suggested methods, the recommendation made by [31, 57] is the yardstick to follow.

2.4.4 Reordering

As stated in the building block hypothesis explained in Chap. 1, the order of genes on a chromosome is critical. The purpose of reordering is to attempt to find the gene order which has the better evolutionary potential. A technique for reordering the positions of genes in the chromosome has been suggested. The order of genes between two randomly chosen positions is inverted within the chromosome. Such a technique is known as Inversion.

For example, consider an integer represented chromosome where two inversion sites, position 3 and position 6, are chosen:

$$1 \quad 2 \quad \underbrace{3 \quad 4 \quad 5 \quad 6}_{\text{inversion region}} \quad 7 \quad 8$$

After the inversion, the order of the genes in the inversion region are reversed. Hence, we have

1 2 6 5 4 3 7 8

[25, 52, 131] combine the features of inversion and crossover into a single operator, e.g. partially matched crossover (PMX), order crossover (OX), and cycle crossover (CX).

2.5 Replacement Scheme

After generating the sub-population (offspring), several representative strategies that can be proposed for old generation replacement exist. In the case of generational replacement, the chromosomes in the current population are completely replaced by the offspring [57]. Therefore, the population with size N will generate N offspring in this strategy.

This strategy may make the best chromosome of the population fail to reproduce offspring in the next generation. So it is usually combined with an elitist strategy where one or a few of the best chromosomes are copied into the succeeding generation. The elitist strategy may increase the speed of domination of a population by a super chromosome, but on balance it appears to improve the performance.

Another modification for generational replacement is that not all of the chromosomes of the subpopulation are used for the next generation. Only a portion of the chromosomes (usually the better will win) are used to replace the chromosomes in the population.

Knowing that a larger number of offspring implies heavier computation in each generation cycle, the other scheme is to generate a small number of offspring. Usually, the worst chromosomes are replaced when new chromosomes are inserted into the population. A direct replacement of the parents by the corresponding offspring may also be adopted. Another way is to replace the eldest chromosomes, which stay in the population for a long time. However, this may cause the same problem as discarding the best chromosome.

CHAPTER 3
INTRINSIC CHARACTERISTICS

Based upon the material that has been described in Chaps. 1 and 2, GA can be used to solve a number of practical engineering problems. Normally, the results obtained are quite good and are considered to be compatible to those derived from other techniques. However, a simply GA has difficulty in tackling complicated, multi-tasking and conflicting problems, and the speed of computation is generally regarded as slow. To enhance the capability of GA for practical uses, the intrinsic characteristics of GA should be further exploited and explored.

There are a number of features that have made GA become a popular tool for engineering applications. As can be understood from the previous chapters, GA is demonstrated as being a very easy-to-understand technique for reaching a solution, and that only a few, simple (sometimes no) mathematical formulations are needed. These are not the only reasons that make GA powerful optimizers, GA are also attractive for the following reasons:

- they are easy to implement in parallel architecture;
- they address multiobjective problems;
- they are capable of handling problem with constraints; and
- they can solve multimodal, non-differentiable, non-continuous or even NP-complete problems [45].

The chapter describes the full details of each item in this category, illustrating how they can be used practically in solving engineering problems.

3.1 Parallel Genetic Algorithm

Considering that the GA already possesses an intrinsic parallelism architecture, in a nutshell, there requires no extra effort to construct a parallel computational framework. Rather, the GA can be fully exploited in its parallel structure to gain the required speed for practical uses.

There are a number of GA-based parallel methods to enhance the computational speed [14, 19]. The methods of parallelization can be classified as

Global, Migration and Diffusion. These categories reflect different ways in which parallelism can be exploited in the GA as well as the nature of the population structure and recombination mechanisms used.

3.1.1 Global GA

Global GA treats the entire population as a single breeding mechanism. This can be implemented on a shared memory multiprocessor or distributed memory computer. On a shared memory multiprocessor, chromosomes are stored in the shared memory. Each processor accesses the particular assigned chromosome and returns the fitness values without any conflicts. It should be noted that there is some synchronization needed between generation to generation. It is necessary to balance the computational load among the processors using a dynamic scheduling algorithm, e.g. guided self-schedule.

On a distributed memory computer, the population can be stored in one processor to simplify the genetic operators. This is based on the farmer-worker architecture, as shown in Fig. 3.1. The farmer processor is responsible for sending chromosomes to the worker processors for the purpose of fitness evaluation. It also collects the result from them, and applies the genetic operators for the production of next generation. One disadvantage of this method is that the worker sits idly while the farmer is handling his job.

Goldberg [50] describes a modification to overcome this potential bottleneck by relaxing the requirement for strict synchronous operation. In such a case, the chromosomes are selected and inserted into the population when the worker processors complete their tasks.

Successful applications of this Global GA approach can be found in [28, 32].

3.1.2 Migration GA

This is another parallel processing mechanism for computing the GA. The migration GA (Coarse Grained Parallel GA) divides the population into a number of sub-populations, each of which is treated as a separate breeding unit under the control of a conventional GA. To encourage the proliferation of good genetic material throughout the whole population, individuals' migration between the sub-populations occurs from time to time. A pseudo code is expressed in Table 3.1.

Figs. 3.2-3.4 show three different topologies in migration. Fig. 3.2 shows the ring migration topology where individuals are transferred between directionally adjacent subpopulations. A similar strategy, known as neighbourhood

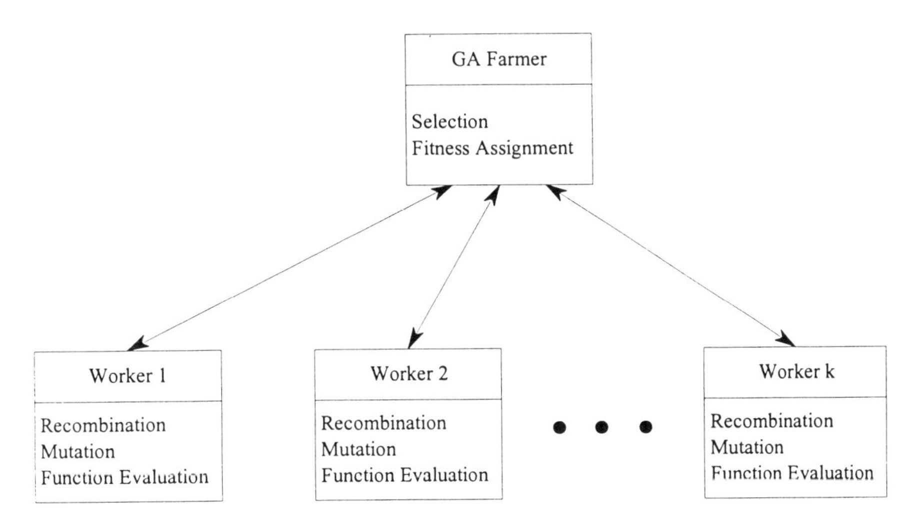

Fig. 3.1. Global GA

Table 3.1. Pseudo Code of Migration GA

```
- Each node (GA)

WHILE not finished
    SEQ
        Selection
        Reproduction
        Evaluation
    PAR
        Send emmigrants
        Receive immigrants
```

migration is shown in Fig. 3.3 where migration can be made between nearest neighbours bidirectionly. The unrestricted migration formulation is depicted on Fig. 3.4 where individuals may migrate from one subpopulation to the other. An appropriate selection strategy should be used to determine the migration process.

The required parameters for a successful migration depend upon:

1. *Migration Rate*
   This governs the number of individuals to be migrated, and
2. *Migration Interval*
   This affects the frequency of migrations.

The values of these parameters are intuitively chosen rather than based on some rigorous scientific analysis. In general, the occurrence of migration is usually set at a predetermined constant interval that is governed by migration intervals.

There are other approaches. [13, 105] introduce migration occurrence once the subpopulation is converged. However, the unknown quantity that determines the right time to migrate remains unsolved. This could cause the migration to occur too early. As a result, the number of correct building blocks in the migrants may be too low to influence a search on the right direction, which eventually wastes the communication resources.

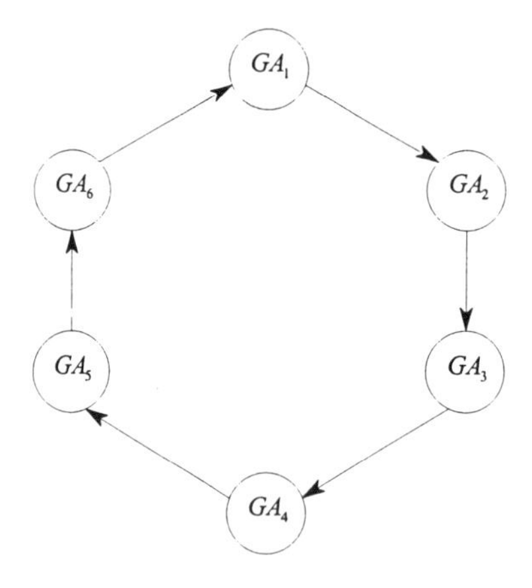

Fig. 3.2. Ring Migration

The topology model of the migration GA is well suited to parallel implementation on Multiple Instruction Multiple data (MIMD) machines. The architecture of hypercubes [22, 150] and rings [55] is commonly used for this purpose. Given the range of possible population topologies and migration paths between them, efficient communication networks should thus

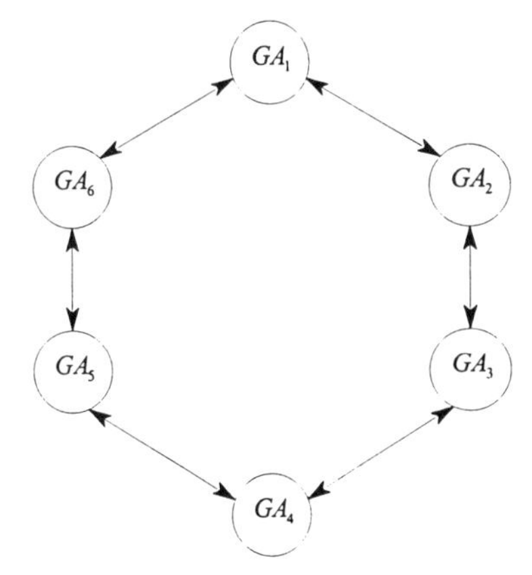

Fig. 3.3. Neighbourhood Migration

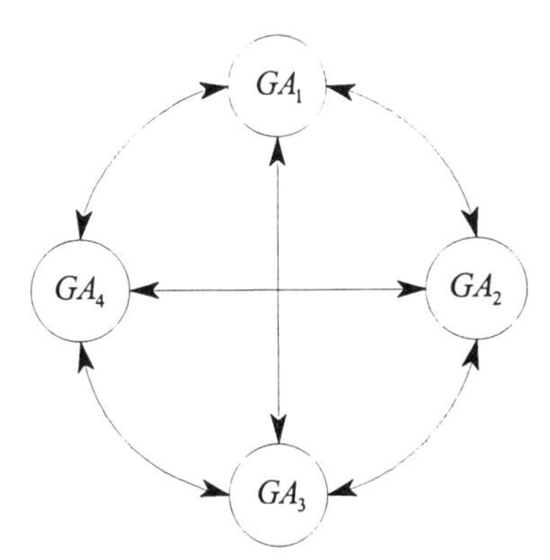

Fig. 3.4. Unrestricted Migration

be possible on most parallel architecture. This applies to small multiprocessor platforms or even the clustering of networked workstations.

3.1.3 Diffusion GA

Apart from the Global and Migration techniques, Diffusion GA (Fine Grained Parallel GA) as indicated in Fig. 3.5, is the architecture can be used as a parallel GA processor. It considers the population as a single continuous structure. Each individual is assigned to a geographic location on the population surface and usually placed in a 2-D grid. This is because of the topology of the processing element in many massively parallel computers that are constructed in this form. Some other different topologies have also been studied in this area [2, 8].

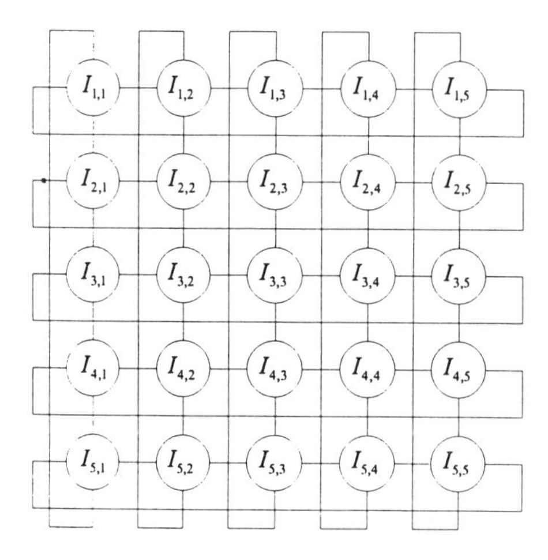

Fig. 3.5. Diffusion GA

The individuals are allowed to breed with individuals contained in a small local neighbourhood. This neighbourhood is usually chosen from immediately adjacent individuals on the population surface and is motivated by the practical communication restrictions of parallel computers. The pseudo code is listed in Table 3.2.

[94] introduced a massive parallel GA architecture with a population distributed with a 2-D mesh topology. Selection and mating were only possible with neighbouring individuals. In addition, [56, 104] introduced an asynchronous parallel GA, ASPARAGOS system. In this configuration, the GA was implemented on a connected ladder network using Transputers with one individual per processor. The practical applications of the Diffusion GA have been reported in [85] to solve a 2-D bin packing problems, and the same technique has been used to tackle a job shop scheduling problem [141].

Table 3.2. Pseudo Code of Diffusion GA

```
- Each node (I_ij)

Initialize
WHILE not finished
    SEQ
        Evaluation
    PAR
        Send self to neighbours
        Receive neighbours
        Select mate
        Reproduce
```

3.2 Multiple Objective

Without any doubt, the GA always has the distinct advantage of being able to solve multiobjective problems that other gradient type of optimizers have failed to meet. Indeed, engineering problems often exist in the class of multiple objectives. Historically, multiple objectives have been combined in an ad hoc manner so that a scalar objective function is formed for the usual linearly combined (weighted sum) functions of the multiple attributes [75, 162]. Another way is by turning the objectives into constraints. There is a lot of work being done using the weighted sums and penalty functions to turn multiobjective problems into single-attribute problems even when a GA is applied. However, a powerful method for searching the multiattribute spaces [50, 41] has been proposed to address these problems.

In this approach, the solution set of a multiobjective optimization problem consists of all those vectors such that their components cannot all be simultaneously improved. This is now known as the concept of Pareto optimality, and the solution set is called the Pareto-optimal set. The Pareto-optimal solutions are also termed as non-dominated, or non-inferior solutions, in which the definition of domination is expressed below:

Definition 3.2.1. *For an n-objective optimization problem, u is dominated by v if*

$$\begin{array}{lll} \forall i = 1, 2, \ldots, n, & & f_i(u) \geq f_i(v) \quad \textit{and} \\ \exists j = 1, 2, \ldots, n, & \textit{such that} & f_i(u) > f_i(v) \end{array}$$

Schaffer proposed a Vector Evaluated GA (VEGA) for finding multiple solutions to multiobjective problems [126]. This was achieved by selecting appropriate fractions of parents according to each of the objectives, separately. However, the population tends to split into species particularly strong in each of the objectives if the Pareto trade-off surface is concave.

Fourman also addressed the multiple objectives in a non-aggregating manner [43]. The selection was performed by comparing pairs of individuals, each pair according to one of the objectives. The objective was randomly selected in each comparison. Similar to VEGA, this corresponds to averaging fitness across fitness components, each component being weighted by the probability of each objective being chosen to decide each tournament.

Pareto-based fitness assignment is the other method, firstly proposed by Goldberg [50]. The idea is to assign equal probability of reproduction to all non-dominated individuals in the population by using non-domination ranking and selection. He also suggested using some kind of niching to keep the GA from converging to a single point. A niching mechanism, such as sharing, would allow the GA to maintain individuals all along the trade-off surface [53]. Ritzel used a similar method, but applied deterministic crowding as the niching mechanism.

Fonseca and Fleming [40] have proposed a slightly different scheme, whereby an individual's rank corresponds to the number of individuals in the current population by which it is dominated.

$$rank(I) = 1 + p \tag{3.1}$$

if I is dominated by other p chromosomes in the population.

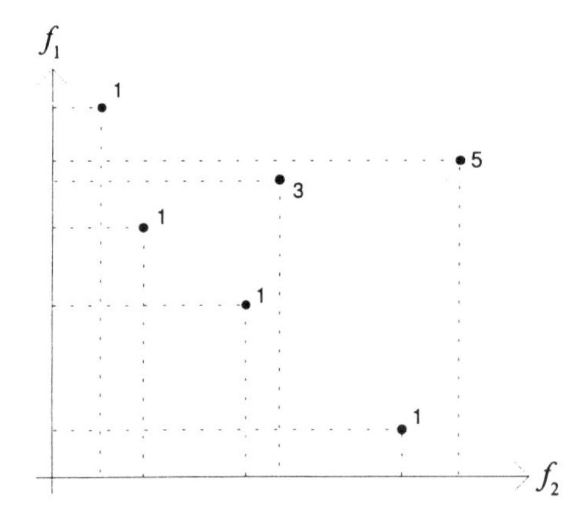

Fig. 3.6. Multiobjective Ranking

Non-dominated individuals in the current population are all within the same rank, see Fig. 3.6, while dominated ones are penalized according to the population density of the corresponding region of the trade-off surface. A theory for setting the niche size is also presented in [40].

This rank-based fitness can also include the goal information in which ranking is based on the preferable rank. Consider two chromosomes I_a and I_b with $F(I_a) = [f_{a,1}, f_{a,2}, \ldots, f_{a,m}]$ and $F(I_b) = [f_{b,1}, f_{b,2}, \ldots, f_{b,m}]$, and the goal vector $V = (v_1, v_2, \ldots, v_m)$ where v_i is the goal for the design objective f_i.

Case 1: $F(I_a)$ meet none of goals
$F(I_a)$ is preferable to $F(I_b)$ $\Leftrightarrow$
$F(I_a)$ is partially less than $F(I_b)$, $F(I_a)\ p\!<\ F(I_b)$, i.e.

$$\forall i = 1, 2, \ldots, m, \quad f_{a,i} \le f_{b,i} \ \wedge\ \exists j = 1, 2, \ldots, m, \quad f_{a,j} < f_{b,j}$$

Case 2: $F(I_a)$ meet all of goals
$F(I_a)$ is preferable to $F(I_b)$ $\Leftrightarrow$

$$F(I_a)\ p\!<\ F(I_b) \vee \sim (F(I_b) \le V)$$

Case 3: $F(I_a)$ partially meet the design goals
Without loss of generality, let

$$\exists k = 1, 2, \ldots, m-1, \quad \forall i = 1, 2, \ldots, k, \quad \forall j = (k+1), (k+2), \ldots, m, \quad (f_{a,i} > v_i) \wedge (f_{a,j} \le v_j)$$

$F(I_a)$ is preferable to $F(I_b)$ $\Leftrightarrow$

$$(f_{a,(1,2,\ldots,k)}\ p\!<\ f_{b,(1,2,\ldots,k)}) \vee (f_{a,(1,2,\ldots,k)} = f_{b,(1,2,\ldots,k)}) \wedge (f_{a,(k+1,k+2,\ldots,m)} p < f_{b,(k+1,k+2\ldots,m)}) \vee \sim (f_{b,(k+1,k+2\ldots,m)} \le v_{(k+1,k+2\ldots,m)})$$

Tournament Selection based on Pareto dominance has also been proposed in [72]. In addition to the two individuals competing in each tournament, a number of other individuals in the population were used to determine whether the competitors were dominated or not. In the case where both competitors were either dominated or non-dominated, the result of the tournament was decided through sharing.

The advantage of Pareto-ranking is that it is blind to the convexity or the non-convexity of the trade-off surface. Although the domination of certain species may still occur if certain regions of the trade-off are simply easier to find than others, Pareto-ranking can eliminate sensitivity to the possible non-convexity of the trade-off surface. Moreover, it rewards good performance in any objective dimension regardless of others. Solutions which exhibit good performance in many, if not all, objective dimensions are more likely to be produced by recombination [90].

Pareto-based ranking correctly assigns all non-dominated individuals the same fitness, but that, on its own, does not guarantee that the Pareto set can be uniformly sampled. When presented with multiple equivalent optima, finite populations tend to converge to only one of these, due to stochastic errors in the selection process. This phenomenon, known as genetic drift, has been observed in natural as well as in artificial evolution, and can also occur in Pareto-based evolutionary optimization.

The additional use of fitness sharing [53] was proposed by Goldberg to prevent genetic drift and to promote the sampling of the whole Pareto set by the population. Fonseca and Fleming [40] implemented fitness sharing in the objective domain and provided a theory for estimating the necessary niche sizes, based on the properties of the Pareto set. Horn and Nafpliotis [72] also arrived at a form of fitness sharing in the objective domain. In addition, they suggested the use of a metric combining of both the objective and the decision variable domains, leading to what was called nested sharing.

The viability of mating is another aspect which becomes relevant as the population distributes itself around multiple regions of optimality. Different regions of the trade-off surface may generally have very different genetic representations, which, to ensure viability, requires mating to happen only locally [50]. So far, mating restriction has been implemented based on the distance between individuals in the objective domain [40, 63].

3.3 Robustness

There are many instances where it is necessary to make the characteristics of the system variables adaptive to dynamic signal behaviour, and ensure that they are capable of sustaining the environmental disturbance. These often require an adaptive algorithm to optimize time-dependent optima which might be difficult to obtain by a conventional GA. When a simple GA is being used, the diversity of the population is quickly eliminated as it seeks out a global optimum. Should the environment change, it is often unable to redirect its search to a different part of the space due to the bias of the chromosomes. To improve the convergency of the standard GA for changing environments, two basic strategies have been developed.

The first strategy expands the memory of the GA in order to build up a repertoire of ready responses to environmental conditions. A typical example in this group is Triallelic representation [54]. Triallelic representation consists of a diploid chromosome and a third allelic structure for deciding dominance.

The random immigrants mechanism [60] and the triggered hypermutation mechanism [20, 21] are grouped as another type of strategy. This approach increases diversity in the population to compensate for the changes encountered in the environment. The random immigrants mechanism is used to replace a fraction of a conventional GA's population, as determined by the replacement rate, with randomly generated chromosomes. It works well in environments where there are occasional, large changes in the location of the optimum.

An adaptive mutation-based mechanism, known as the triggered hypermutation mechanism, has been developed to adapt to the environmental change. The mechanism temporarily increases the mutation rate to a high value (hypermutation rate) whenever the best time-average performance of the population deteriorates.

A simulation has been conducted for illustrating the response of the GA to environmental changes. The task was to locate the global maximum peak, numerically set to 60, for the landscape depicted in Fig. 1.13. It had two-variable functions and each variable was represented in 16 bits. In other words, each population member was 32 bits long. The other parameter settings of the GA are tabulated in Table 3.3.

Table 3.3. Parameter Settings of Conventional GA

Representation	16-bit per variable (total 32 bit)
Population size	100
Generation gap	0.8
Fitness assignment	ranking
Selection	roulette wheel selection
Crossover	one-point crossover
Crossover rate	0.6
Mutation	bit mutation
Mutation rate	0.001

Environmental changes were introduced to testify as to the robustness of the conventional GA and the mechanisms, Random Immigrant Mechanism and Hypermutation. The changes were as follows:

1. linear translation of all of the hills in the first 50 generations. The hill location was increased by one step in both dimensions after five generations. Each dimension's rate of change was specified independently, so that one dimension might increase while another was decreased; and

2. relocation of the maximum hill randomly every 20 generations in the period of 50-150 generations, while keeping the remainder of the landscape fixed.

Table 3.4 summarizes the simulation results.

Table 3.4. Simulation Results

	Parameter setting	Result
Conventional GA	Table 3.3	Fig. 3.7
Random Immigrant	Replacement rate = 0.3	Fig. 3.8
Hypermutation	Hypermutation rate = 0.1	Fig. 3.9

In Fig. 3.7, it can be observed that the conventional GA is unable to relocate the global maximum. As explained before, this is due to the lack of population diversity. It can be revealed by comparing the average fitness and the best fitness in which these two values are approximately the same.

The Random Immigrant Mechanism and Hypermutation performed well in the experiment. However, it is worth noting that Hypermutation is triggered only when a change in the environment decreases the value of the best of the current population. In some situations, for example, if a new optimum, of say 70, occurred in another area with all the others remaining unaltered, there would be no triggers and the Hypermutation would be unlikely to detect the change.

Moreover, in a noisy condition, it is difficult to determine a change of environment. Statistical process control is hence proposed [143] to monitor the best performance of the population so that the GA-based optimization system adapts to the continuous, time-dependent nonstationary environment. The actual implementation can be referred to in the case study described in Chap. 4.

3.4 Multimodal

The other attribute of the GA is its capability for solving multimodal problems. Three factors [50] contribute to its ability to locate a global maximum:

- searching from a population of solutions, not a single one
- using fitness information, and not derivatives or other auxiliary information; and
- using randomized operators, not deterministic ones.

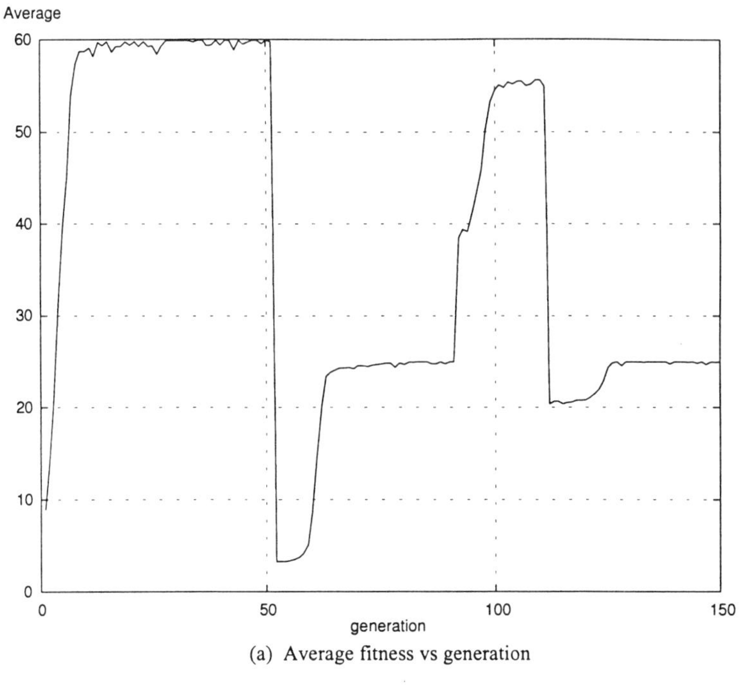

(a) Average fitness vs generation

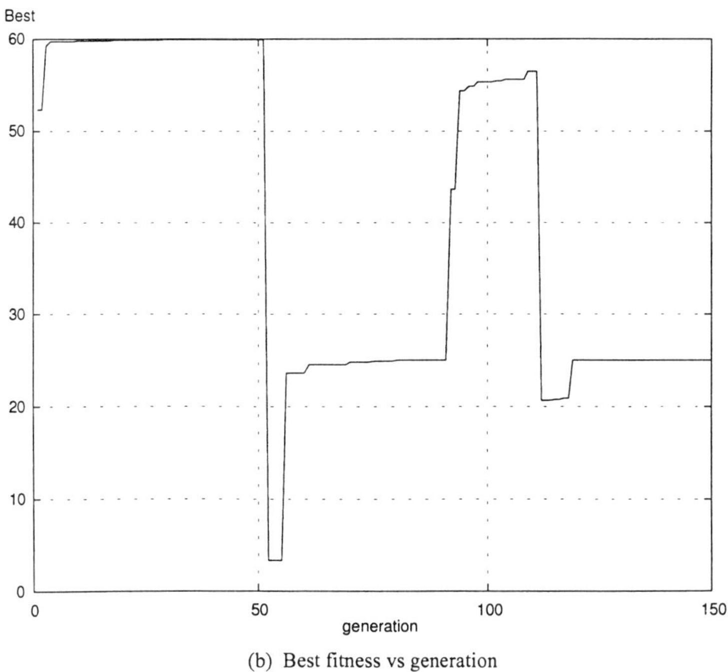

(b) Best fitness vs generation

Fig. 3.7. Conventional GA

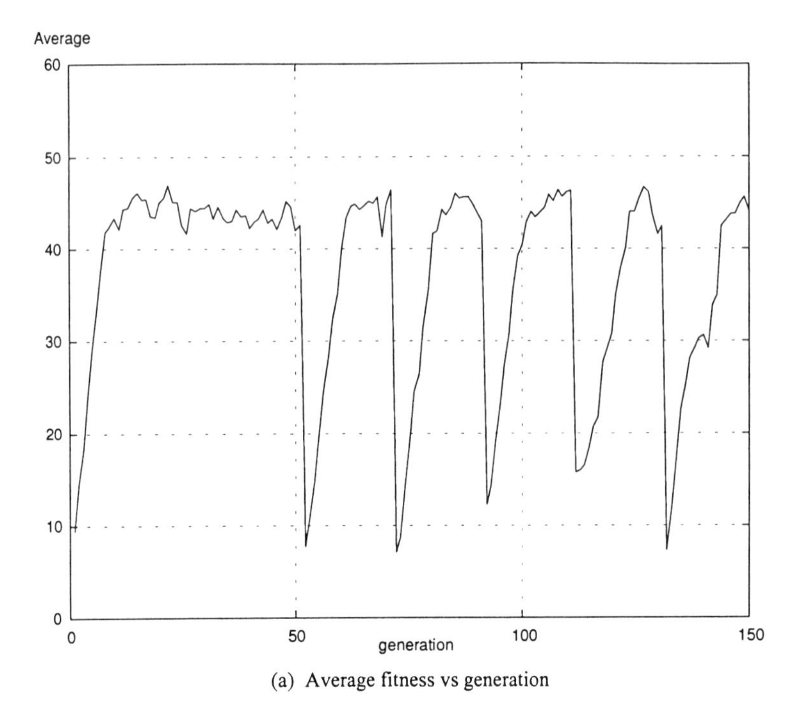

(a) Average fitness vs generation

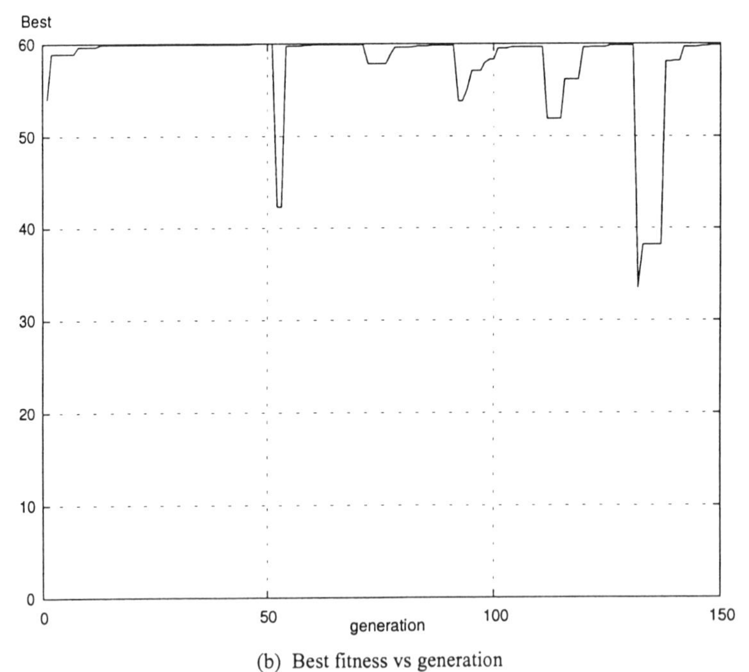

(b) Best fitness vs generation

Fig. 3.8. Random Immigrant Mechanism

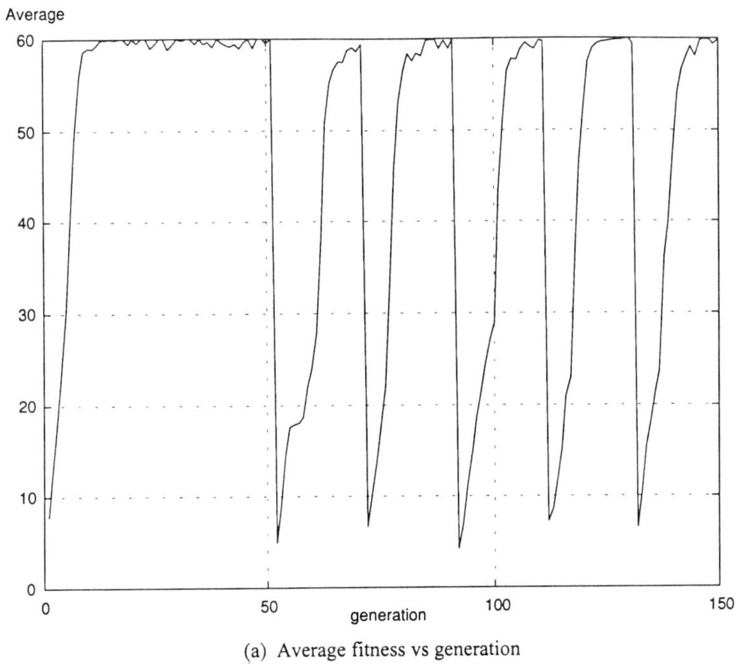

(a) Average fitness vs generation

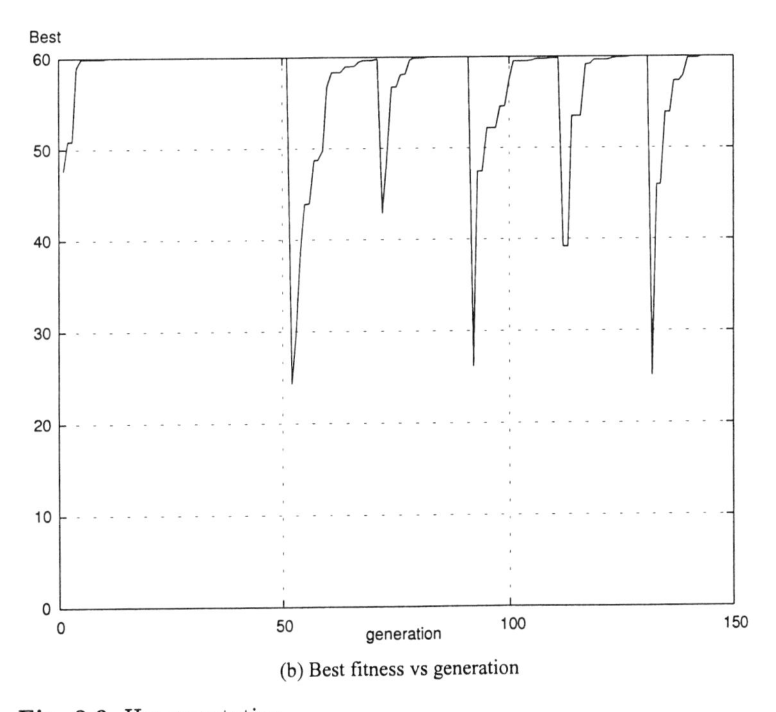

(b) Best fitness vs generation

Fig. 3.9. Hypermutation

However, there is no guarantee that the global optimal point will be obtained by using GA although there is the tendency for this to occur. The possibility of success is reduced if there is a loss of population diversity. As the GA has a tendency to seek a sub-optimal point, the population may converge towards this value which leads to premature convergence. The global optimal solution is only obtained via the exploration of mutation in the genetic operations. Such a phenomenon is known as a Genetic Drift [12] and this situation occurs easily with a small size of population.

A number of techniques have been proposed to limit the effect of genetic drift and maintain the population diversity. These include Preselection [91], Crowding [30, 50] and Fitness Sharing [42].

3.5 Constraints

In the process of optimization, the problem of constraint is often encountered. This obstacle is not always handled properly by the conventional, but mathematically governed optimization techniques. By contrast, constraints present no problems to the GA and various methods can be used in this area.

3.5.1 Searching Domain

It is possible to embed the condition of constraints in the system by confining the searching space of a chromosome. This approach guarantees that all chromosomes are valid and that the constraint will not be violated. A typical example is to limit the searching domain of the coefficients of a digital lattice filter design in the range of $+1$ to -1 whose pole locations will be confined within the unit circle for stability. This method of solving the constraint problem requires no additional computing power, and all chromosomes created are regarded as potential solutions to the problem.

3.5.2 Repair Mechanism

The repair mechanism is a direct analogy to the DNA repair systems in the cell. In DNA, there are numerous repair systems for cell survival and the diverse sources of DNA damage. These systems may cure mismatching problem, abnormal bases, etc. This process can be emulated by the GA to solve the constraint problem. If any condition of the constraint is violated by a chromosome, the chromosome will be *"corrected"* so that it becomes valid.

This can be achieved by modifying some genes randomly within the valid solution space, or backtracking toward its parents' genetic material

[81]. However, this method is rather computationally intensive. It behaves in the same way as DNA, where repairing is extraordinarily inefficient in an energetic sense. Nevertheless, it is a worthy direction to approach since the fitness evaluation is normally the most time-consuming process. The repair mechanism can ensure that each chromosome undertaking the fitness evaluation is a potential solution for the constrained problem.

3.5.3 Penalty Scheme

Another approach for handling constraints is to set up a penalty scheme for invalid chromosomes such that they become low performers. The constrained problem is then transformed to an unconstrained condition by associating the penalty with all the constraint violations. This can be done by including a penalty to adjust the optimized objective function.

Consider Eqn. 3.2 as the original objective function to be optimized,

$$f(x_1, x_2, \ldots, x_n) \tag{3.2}$$

To comprise a penalty scheme within the objective function, Eqn. 3.2 becomes

$$f(x_1, x_2, \ldots, x_n) + \delta \sum_{i=1}^{m} \varphi_i \tag{3.3}$$

where m is the total number of constraints; δ is a penalty coefficient which is negative for maximization and positive for minimization problems; and φ_i is a penalty related to the i-th constraint (i=1,2,...,m).

The penalty scheme has two distinct characteristics:

- some vital information may be thrown away; and
- a small violation of a constraint may qualify if it produces a large payoff in other areas.

However, an appropriate penalty function is not easy to come by and affects the efficiency of the genetic search [119]. Moreover, computing time is wasted in evaluating the invalid chromosome, especially when the problem is one in which the constraints are likely to be violated.

3.5.4 Specialized Genetic Operations

Michalewicz introduced another direction for handling the constrained numerical problem based on his software package GENOCOP (Genetic Algorithm for Numerical Optimization with Linear Constraints) [101]. The floating point representation was used and the crossover and mutation operations were specifically designed so that the newly generated offspring were included in the solution space.

Mutation. For a convex space S, every point $s_0 \in S$ and any line p such that $s_0 \in p$, p intersects the boundaries of S at precisely two points, denoted by $l_p^{s_0}$ and $u_p^{s_0}$.

3 different mutation operations are designed based on this characteristic.

1. For uniform mutation, the mutating gene v_k which is the k-th component of chromosome s_t is a random value from the range $[l_{(k)}^{s_t}, u_{(k)}^{s_t}]$ where $l_{(k)}^{s_t}$ and $u_{(k)}^{s_t}$ are the k-th components of the vectors $l_p^{s_t}$ and $u_p^{s_t}$, respectively and $l_{(k)}^{s_t} \leq u_{(k)}^{s_t}$.
2. For boundary mutation, the mutating gene v_k is either $l_{(k)}^{s_t}$ or $u_{(k)}^{s_t}$ with equal probability.
3. For non-uniform mutation, the mutating gene v_k is modified as

$$v_k = \begin{cases} v_k + \Delta(t, u_{(k)}^{s_t} - v_k) & \text{if a random digit is 0} \\ v_k - \Delta(t, v_k - l_{(k)}^{s_t}) & \text{if a random digit is 1} \end{cases}$$

Crossover. [101] designed three crossovers based on another characteristic of the convex space:

For any two points s_1 and s_2 in the solution space S, the linear combination $a \cdot s_1 + (1-a) \cdot s_2$, where $a \in [0,1]$, is a point in S.

Consider two chromosomes $s_v^t = [v_1, \ldots, v_m]$ and $s_w^t = [w_1, \ldots, w_m]$ crossing after k-th position,

1. For single crossover, the resulting offspring are

$$\begin{aligned} s_v^{t+1} &= [v_1, \ldots, v_k, w_{k+1} \cdot a + v_{k+1} \cdot (1-a), \\ &\qquad \ldots, w_m \cdot a + v_m \cdot (1-a)] \\ s_w^{t+1} &= [w_1, \ldots, w_k, v_{k+1} \cdot a + w_{k+1} \cdot (1-a), \\ &\qquad \ldots, v_m \cdot a + w_m \cdot (1-a)] \end{aligned}$$

2. For single arithmetical crossover, the resulting offspring are

$$\begin{aligned} s_v^{t+1} &= [v_1, \ldots, v_{k-1}, w_k \cdot a + v_k \cdot (1-a), v_{k+1}, \ldots, v_m] \\ s_w^{t+1} &= [w_1, \ldots, w_{k-1}, v_k \cdot a + w_k \cdot (1-a), w_{k+1}, \ldots, w_m] \end{aligned}$$

3. For whole arithmetical crossover, the resulting offspring are

$$s_v^{t+1} = a \cdot s_w^t + (1-a) \cdot s_v^t$$
$$s_w^{t+1} = a \cdot s_v^t + (1-a) \cdot s_w^t$$

This is an effective method for handling constraints for numerical problems, but there is a limitation in solving non-numerical constraints such as the topological constraints found in networking [109].

CHAPTER 4
ADVANCED GA APPLICATIONS

In essence, the materials that are described in the previous chapters are sufficient for most engineering applications. Not only are they covered in depth, but a wide spectrum of GA knowledge for tackling problems that are difficult to solve by other techniques is given. To demonstrate each special characteristic of the GA individually is not easy, and sometimes these have been combined together in order to obtain the required solutions. Therefore, the purpose of this chapter is to consider three special case studies in which the main features of the GA are outlined. It is also hoped that the advantages of using GA for engineering designs can be highlighted and used as a reference for designing future systems.

4.1 Case Study 1: GA in Time Delay Estimation

In this case study, a GA is applied to tackle an on-line time-delay estimation (TDE) problem. TDE can be found in many signal processing applications such as sonar, radar, noise cancellation, etc. It is usually solved by changing the one-dimensional delay problem into a multidimensional problem with finite impulse response (FIR) filter modelling [15, 118, 165]. Adaptive filtering techniques [66, 160] have been successfully applied in this area relying on the unimodal property of FIR error surface. Due to the GA's ability in locating global optima on a multimodal error surface, direct estimates of delay as well as the gain parameter are possible.

4.1.1 Problem Formulation

Estimation of the time delay between signals received from two sensors is considered. Eqn. 4.1 represents the conventional discrete time delay model.

$$\begin{aligned} x(kT) &= s(kT) + v_1(kT) \quad \text{and} \\ y(kT) &= \alpha s(kT - D) + v_2(kT) \end{aligned} \tag{4.1}$$

where T is the sampling period; $x(kT)$ and $y(kT)$ are two observed signals at sampled time kT; $s(kT)$ is the transmitted source, white signal with power σ_s^2; D is the delay to be estimated; α is a gain factor between the sensors;

and $v_1(kT)$ and $v_2(kT)$ are zero-mean stationary noise processes with power σ_n^2, assumed to be uncorrelated with each as well as $s(kT)$.

In principle, the searching domain of time delay includes infinite numbers of values of D and α. In practice, these numbers are limited by the knowledge of the delay and gain range and by the desired resolution. Moreover, the resolution of delay is often much finer than the sampling interval T. Assuming that the measured signals are band-limited with a frequency range $(-\omega, \omega)$ which leads to $T \leq (2\omega)^{-1}$, the signal $x(kT)$ can be interpolated in the form of sinc function [15]. Therefore, the delayed version $x(kT - D)$ can be approximated by

$$x(kT - D) \approx \sum_{i=-L}^{L} sinc(i - D)x(k - i) \tag{4.2}$$

The estimation error, $e(k)$, is defined as

$$e(k) = y(k) - AX(k) \tag{4.3}$$

where $X(k) = \begin{bmatrix} x(k+L) & x(k+L-1) & \cdots & x(k-L) \end{bmatrix}^T$ is the input vector, and A is the vector to be optimized.

The optimal solution of A ($\tilde{A}$) for minimum mean square error (MMSE) criterion is expressed as

$$\tilde{A} = \frac{\alpha\sigma_s^2}{\sigma_s^2 + \sigma_n^2} [\ sinc(-L - D) \quad sinc(-L + 1 - D) \quad \cdots \quad sinc(L - D)\ ] \tag{4.4}$$

which is only dependent on α and D. Hence, the problem is now reduced to two parameters $(g, \hat{D})$:

$$g = \frac{\alpha SNR}{1 + SNR} \quad \text{and} \quad \hat{D} = D \tag{4.5}$$

where SNR is the signal to noise ratio.

The optimal set of $(g, \hat{D})$ lies upon the multimodal mean square error surface which might not be easily obtainable using gradient searching methods. Hence, optimization of A vector leads to the result of an FIR filter. However, such a filtering model often causes estimation noise problem as more parameters are required for estimation. To combat this difficulty, the Constrained LMS Algorithm [133] to search for $\hat{D}$ directly with the assumption that the initial guess $\hat{D}(0)$ is within the range of $D \pm 1$ has been proposed.

4.1.2 Genetic Approach

To directly estimate the optimal set of $(g, \hat{D})$ and avoid any restriction of the initialization, a GA can be introduced. The associated delay and gain in this case, directly represented by a binary string, i.e. chromosome, without modelling of the delay is a significant advantage. This chromosome is defined in Eqn 4.6 and its structure is depicted in Fig. 4.1.

$$I = (b_1, b_2, \ldots, b_{32}) \tag{4.6}$$

where $b_i \in B = 0, 1$

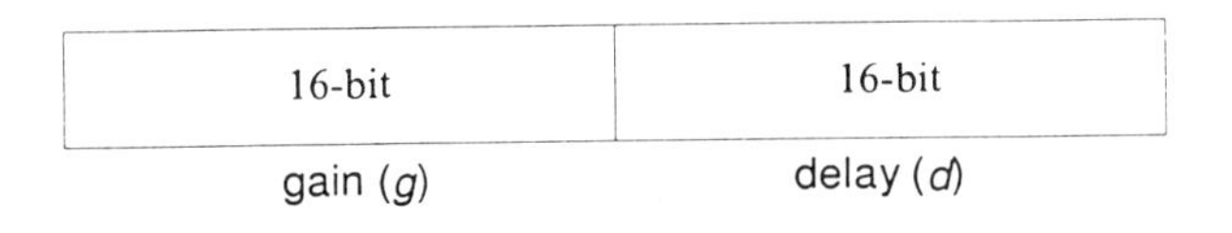

Fig. 4.1. Chromosome Structure

Consider a transformation from a binary to a real number, $\Lambda : B^{16} \times B^{16} \longrightarrow \Omega \subseteq \Re \times \Re$ converting the chromosome bit string $I = (b_1, b_2, \ldots, b_{32})$ to phenotype values $(g, \hat{D})$. The overall goal of GA for time delay problems is to obtain the chromosome $I^* \in B^{16} \times B^{16}$ with the phenotype value $\Lambda(I^*)$such that

$$\Psi(\Lambda(I)) \geq \Psi(\Lambda(I^*)) = \Psi^* \qquad \forall I \in \Omega \tag{4.7}$$

where $\Psi : \Omega \longrightarrow \Re^+$ is the average square difference function expressing as $\Psi(g, \hat{D}) = \frac{1}{N} \sum_{k=k_0}^{k_0+N-1} \left[y(k) - g \sum_{i=-L}^{L} sinc(i - \hat{D}) x(k-i \right]^2$; Ω is the searching domain for $(g, \hat{D})$; N is the estimation window size; k_0 is the starting sample of the window; Ψ is called a global minimum, and $\Lambda(I^*) = (g^*, \hat{D}^*)$ is the minimum location in the searching space Ω.

In order to apply a GA for real time TDE applications as stated in Eqns. 4.1 to 4.7, the problem on noise immunity and robustness should be addressed.

To tackle the system noise immunity problem, that is when the noise level is comparative to the actual signal, i.e. a low SNR value, the obtained result is not steady and often yields high variance. In order to improve the noise immunity, the phenotype values for the best chromosomes obtained in the generations are stored in particular memory locations. The mean of these phenotype values is applied for real world interaction instead of the phenotype value of the current best chromosome.

Furthermore, in order to reach a global optimum, a strong selection mechanism and a small mutation rate are often adopted during the GA operation, in order to contract diversity from the population through the searching process. Should the environmental condition be changed, the conventional GA is unable to redirect its search to the new optimum speedily. In this system, a statistical monitoring mechanism is necessary and implemented. By monitoring the variance and the mean of the output data, a change of environment can be detected. It is assumed that the process is subject only to its natural variability and remains in a state of statistical control unless a special event occurs. If an observation exceeds the threshold, a significant deviation from the normal operation is deemed to have occurred. The GA operations have to be readjusted to adapt changes.

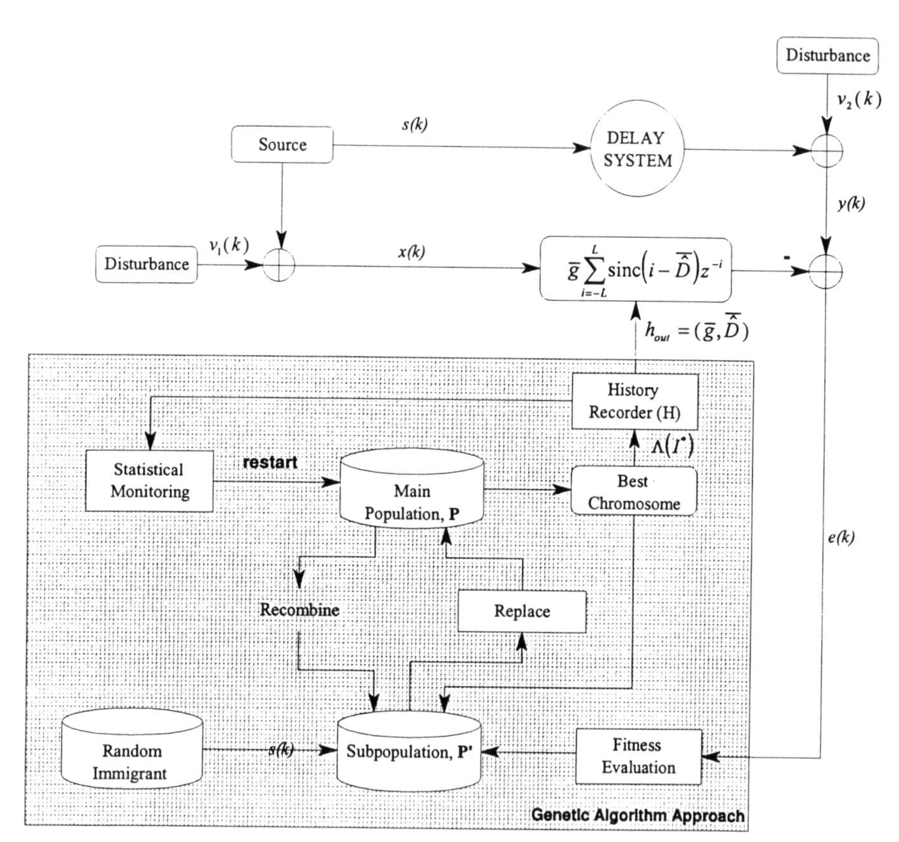

Fig. 4.2. Block Diagram of Genetic-TDE System

By taking all these considerations into account, a genetic-based time delay estimation system is thus proposed and schematically depicted in Fig. 4.2. Basically the system has three modes of operation, namely, Initialization,

Learning and Monitoring. The procedures for each mode are described as follows:

Initialization Mode. The activation of the TDE system should first be started with initialization procedures as follows:

1. Generate the chromosomes, I, as stated in Eqn. 4.6 for the main population *(P)* with uniform distribution;
2. Evaluate the objective value, Ψ, as in Eqn. 4.7 for each chromosome from population P

Learning Mode. Once the initialization mode has been completed, the GA operations can proceed. This Learning Mode enables chromosomes to be improved while the system is in operation. The population is updated recursively, and always supplies the mean of the past best phenotype values for real world operation. The learning cycle is depicted in Fig. 4.3.

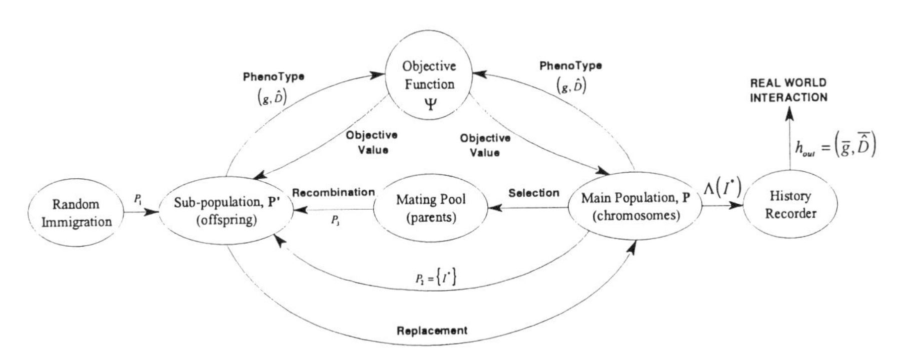

Fig. 4.3. Learning Cycle

The following GA operations are used to ensure the speed of computation as well as to improve the noise immunity within the learning cycle.

Parent Selection. Parent Selection is a routine to emulate the survival-of-the-fittest mechanism of nature. It is expected that a better chromosome will receive a higher number of offspring and thus have a higher chance of surviving in the subsequent generation. The chromosomes in the population pool are selected for the generation of new chromosomes (offspring) by the stochastic universal sampling (SUS) method [7].

Formulation of Sub-population (P'). The new chromosomes in the sub-population (P') are generated by combining three different sets:

$$P'_k = P_{1,k} \cap P_{2,k} \cap P_{3,k} \tag{4.8}$$

where $P_{1,k}$ is the set of chromosomes to be generated via random immigration; $P_{2,k} = I^*_k$ contains the best chromosome in the current main population; and $P_{3,k}$ is the set of chromosomes generated through a recombination process.

Genetic Operation. 2-point crossover and bit mutation are applied.

Replacement. In order that a best chromosome is available for real-world interaction, the main population is discarded and replaced by the newly-obtained subpopulation (P') in the next generation. That is when

$$P_{k+1} = P'_k \tag{4.9}$$

Fitness Assignment. To prevent premature convergence, the fitness value of the chromosome, $f(I,k)$, is assigned by a linear ranking scheme [6].

History Recorder. This scheme is necessary in order to maintain the best chromosome of the current population for real-time interaction. The History Recorder H, is structured as

$$\begin{aligned} h(i) &= h(i+1) \quad \forall\, i \in (0,N) \\ h(N) &= \Lambda(I_k^*) \end{aligned} \tag{4.10}$$

where I_k^* is the best chromosome in the k-th generation; N is the maximum size of the History Recorder; and $h(i)$ is the i-th element in History Recorder.

To further improve the system noise immunity capability, the mean value of the elements in the History Recorder, (h_{out}), for real world operation is adopted:

$$h_{out} = \frac{1}{N}\sum_{i=1}^{N} h(i) \tag{4.11}$$

Monitoring Mode. Under normal circumstances, the TDE system should properly operate in a learning mode. Its performance should thus be improved and enhanced further as time continues. However, this mode of operation must be guaranteed once the system is disturbed. Therefore, in order to increase the system's robustness, since it has the capacity to reject external disturbance, a monitoring scheme is set-up in parallel for the on-line checking of the environment. This can be established by using the statistical output value to provide the necessary information about a change of situation. The scheme is formulated by considering the set $S \subset H$, which consists of n output values from current sample (k) as to the past sample ($k-n-1$), and the system is thus considered as being converged or unchanged when the standard deviation is less than a particular value σ_k. This value can be obtained either experimentally or where a prior knowledge is available.

$$\sigma(S) < \sigma_k \tag{4.12}$$

The system should only be re-set when

$$\begin{aligned} |\hat{D} - \bar{D}| &> \delta_D \quad \text{or} \\ |g - \bar{g}| &> \delta_g \end{aligned} \tag{4.13}$$

where $(g, \hat{D})$ is the current output value; $(\bar{g}, \bar{D})$ is the mean value after convergence; δ_D, $\delta_g$are the threshold values which can be determined as the multiple of the σ_k or the expected accuracy of the system.

4.1.3 Results

To illustrate the effectiveness of the proposed method for TDE, both time invariant and variant delay cases were studied. In the case of invariant time delay, only the estimation of the time delay element was considered, whereas for time variant delay, the introduction of both gain and delay parameters were given. To further ensure the tracking capability and robustness of the system, which was generally governed by the monitoring scheme, the previous experimental run was repeated again while the monitoring mode was being switched off. A direct comparison of the results obtained from both cases was made to assess the function of this mode.

Invariant Time Delay. For the estimation of time-invariant time delay, the gain was fixed to unity and the sampling time $T = 1$. Since only the time delay was to be estimated, the chromosome was formulated as a 16-bit string which solely represented the delay, see Fig. 4.6. The parameters used for simulation runs were $\sigma_s^2 = 1$, $L = 21$, $D = 1.7$ and $N_p = 8$. The genetic operations consisted of a two-point crossover with an operation rate $p_c = 0.8$ and a mutation with a rate $p_m = 0.01$. Objective Function defined in Eqn 4.7 was evaluated while the window size was arbitrarily limited to five samples. Figs. 4.4 and 4.5 show the comparison with LMSTDE [160] and the Constrained LMS Algorithm [133] whose formulations are stated in Appendix A and B for both noiseless and noisy condition ($SNR = 0dB$). For the noiseless situation, a direct output of the current best chromosome was applied.

The mean and variance of different algorithms are compared in Table 4.1. It can be observed that the proposed method provides a better delay estimation as compared with the two other methods.

Table 4.1. Statistical Comparison of Different Algorithms

Algorithms	Mean	Variance (10^{-4})
Proposed Algorithm	1.6982	0.569
Constrained LMS Algorithm	1.6827	7.714
LMSTDE	1.7045	2.584

Variant Time Delay. In order to demonstrate the robustness of the proposed system, an estimation of time variant delay is also considered. The sequences $s(kT)$, $v_1(kT)$ and $v_2(kT)$ were generated by a random

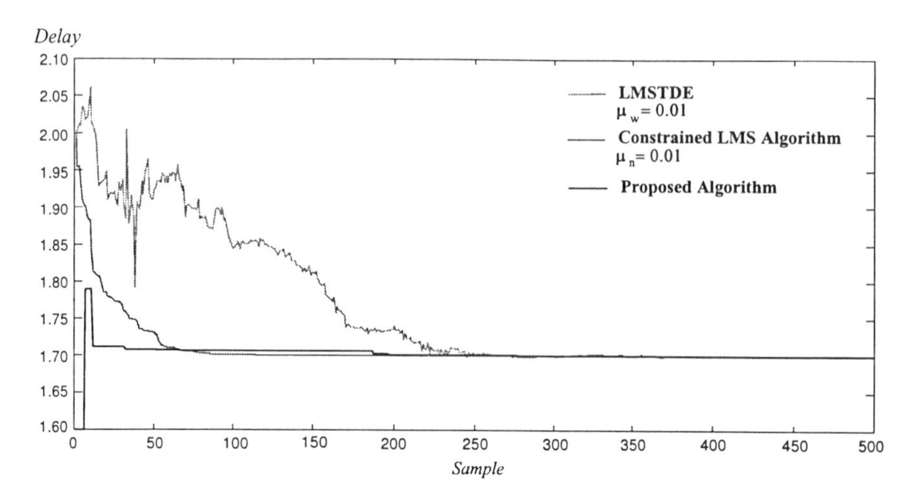

Fig. 4.4. Comparison of Different Algorithms (Noiseless Condition)

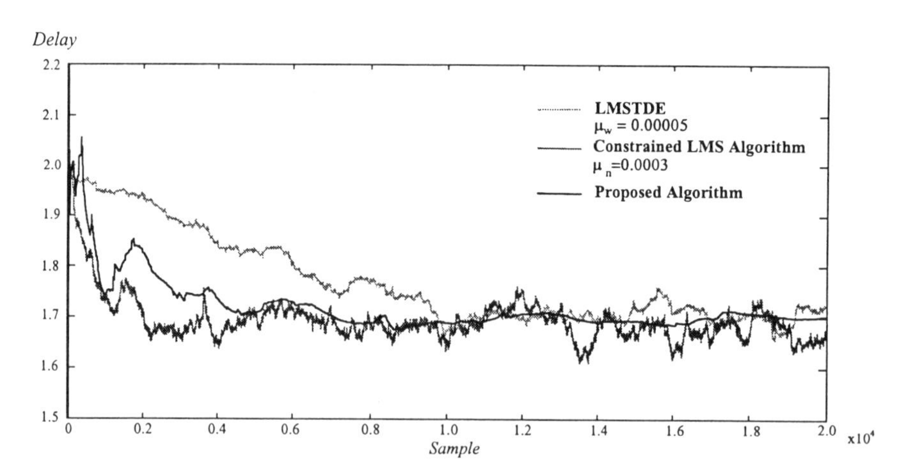

Fig. 4.5. Comparison of Different Algorithms ($SNR = 0dB$)

number generator of Gaussian distribution with the signal power σ_s^2 set to unity. The tracking ability of the proposed method for noiseless condition is demonstrated in Fig. 4.6. A step change on the parameters of gain and delay for the TDE system occurred in every 1,000 samples.

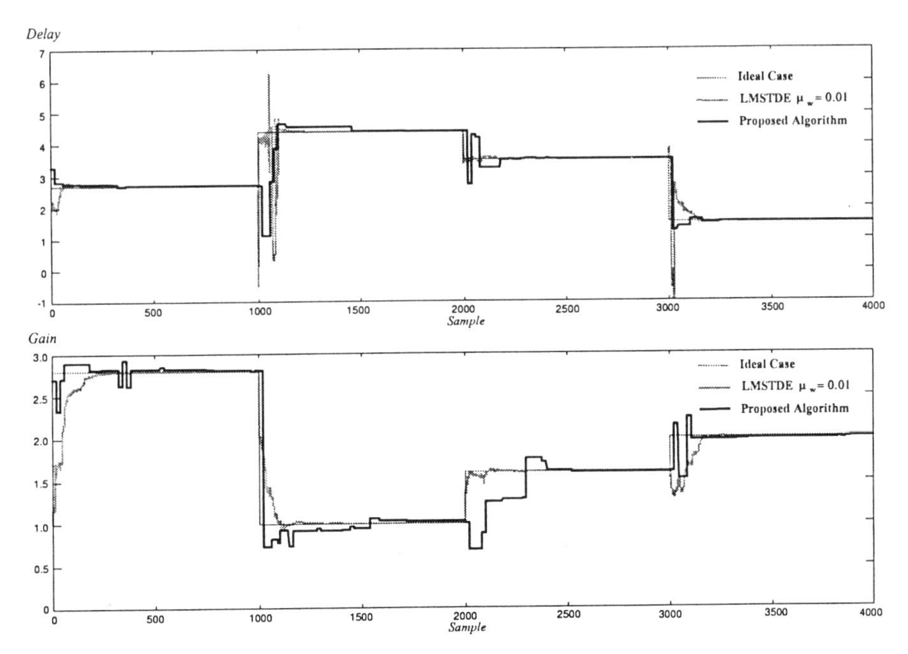

Fig. 4.6. Time Delay Tracking by Different Algorithms (Noiseless Condition)

For the noiseless case, the current best chromosome is considered as the output. It should be noted that the Constrained LMS Algorithm is not applicable for time variant delay problems, as an initial guess is not easily determined within the required limit, $D \pm 1$. Hence, comparison is only possible via the traditional LMSTDE method.

Fig. 4.7 shows the results obtained for the case when SNR=0dB is set. Both variation of gain and delay changes remained unchanged as from previous cases. It can be seen that the system behaves well under extremely noisy conditions.

To illustrate the effectiveness of the monitoring system, the previous simulation run was repeated once again, but the monitoring mode was switched off. The tracking ability of the system was poor as indicated in Fig. 4.8. It can be observed that the system reacts very slowly to the changing environment in the absence of the monitoring scheme.

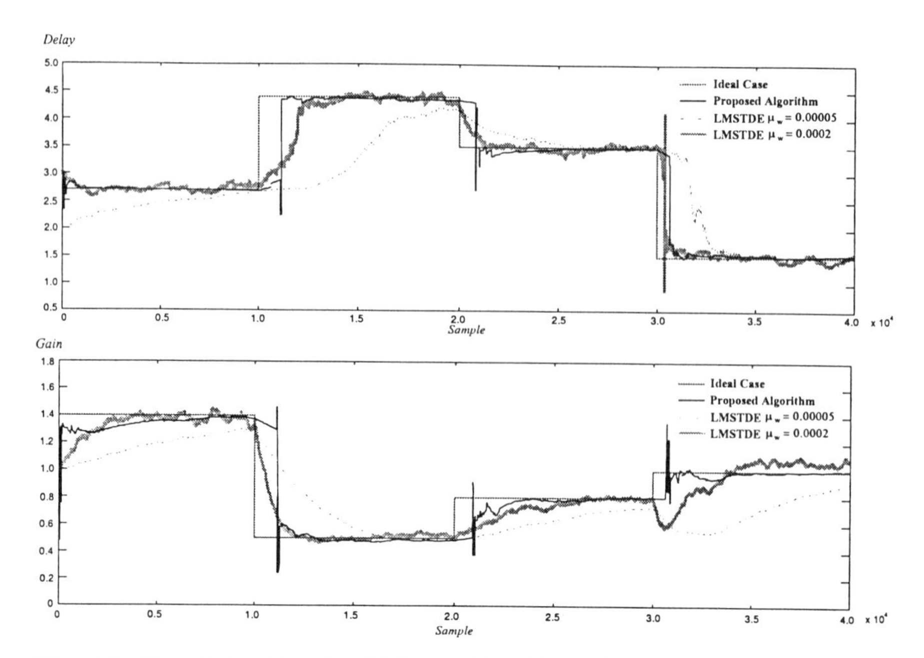

Fig. 4.7. Time Delay Tracking Different Algorithms (SNR = 0dB)

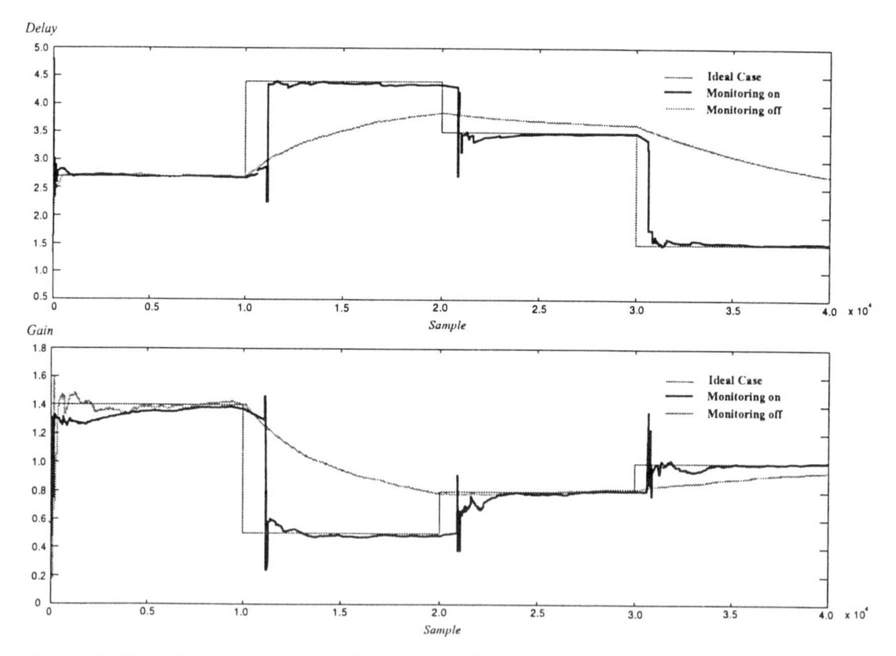

Fig. 4.8. Slow Tracking When Statistical Monitoring Mechanism Off

4.2 Case Study 2: GA in Active Noise Control

The concept of noise reduction using active noise control (ANC) has been with us for a long time. The very first patent design by Leug [89] was established in 1936. ANC is a technique that uses secondary acoustic sources to generate the anti-phase sound waves that are necessary for cancelling undesired noise. It works on the basic principle of destructive interference of sound fields. A number of ANC systems have been developed for different applications [35]. The more notable systems are: ANC systems for power transformers [11, 17], duct systems [64, 111], and systems for vehicle or flight cabins [34, 136]. The main contribution to the success of these systems is the use of parameter identification techniques for the estimation of the noise dynamics and the development of adaptive optimal control laws to govern the corresponding anti-phase acoustic signals for final noise cancellation. An effective ANC system must be able to adjust the controller to activate the secondary sources in such a way that the resultant noise received by the error sensor is reduced to a minimum.

Therefore, speedy and accurate estimates of the acoustic dynamics are the keys elements to the success of ANC systems. This generally involves the determination of the correct estimates of the amplitude and phase of noise signals. Thus far, the well known gradient climbing type [46, 160] of optimization scheme has normally been adopted for this purpose. However, this process is not without its shortcomings. The most noticeable adverse effect is the inherent phenomenon of being trapped in the multimodal surface during the process of optimization. Another problem is the structure of parallelism for high speed computation which is considered an essential technique for the development of complex and multichannel systems.

In this chapter, the ANC system adopts the intrinsic properties of the GA. A number of new schemes are proposed in this development. These include the modelling of acoustic path dynamics, the parallelism of computation and a noticeable multiobjective design for sound field optimization. An advanced ANC design that uses dedicated hardware for GA computation is also recommended. The advantage of developing this architecture is its simplicity in system integration, and the fact that an expensive DSP processor is no longer required for calculating the time consuming numerical values.

4.2.1 Problem Formulation

To configure an ANC system based on a GA formulation, the overall feed-forward ANC system can be seen from Fig. 4.9. The aim is to minimize the sound field at "D" via "C" for generating the anti-phase noise signal to counteract the noise produced by "A". This is only possible if "B", the detector (microphone) is placed adjacent to "A" and sends its output signal to the controller $C(z)$ to adjust the outgoing noise by the actuators (loudspeakers). An optimal noise reduction performance would only be on hand if another microphone(s) is located at "D" and feeds its output signal to the controller $C(z)$ via some intelligent optimization routines.

In practice, other adverse conditions can be contributed by the locations of the microphones and the loudspeakers. A positive acoustic feedback path always exists due to a contaminated secondary signal sensed by the microphone at "B", i.e. through location "C" to "B". This path introduces an additional dynamic into the overall noise dynamics. The other transfer function relating to the loudspeaker and the error sensor, i.e. through location "C" to "D", is non-unity, time-varying and unknown. A slight positional mismatch between the loudspeakers and the error sensor will cause significant deterioration in the performance of ANC systems. Therefore, it is necessary to take these problems into account while the identification procedure is taking place. Generally, this process can be completed by the well known filtered-x Least Mean Squares [36, 111].

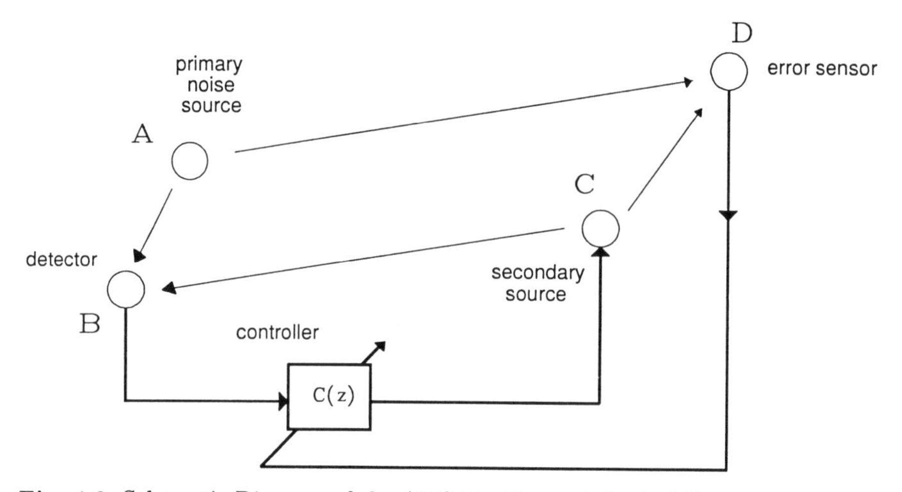

Fig. 4.9. Schematic Diagram of the ANC Feedforward Control System

Mathematically, the problem can be formulated according to the general block diagram of the ANC system as depicted in Fig. 4.10. In order that $o(k)$ receives minimal energy due to $s(k)$ propagation, $u(k)$ has to generate appropriate anti-phase signals, such that

$$o(k) = H_3(z)s(k) + H_4(z)u(k) \tag{4.14}$$

$$m(k) = H_1(z)s(k) + H_2(z)u(k) \tag{4.15}$$

Substituting Eqn. 4.15 into Eqn. 4.14, we have,

$$o(k) = H_3(k)\frac{m(k) - H_2(z)u(k)}{H_1(z)} + H_4(z)u(k) \tag{4.16}$$

Eqn. 4.16 is a global formulation of all ANC configurations including its application in a 3-D propagation medium.

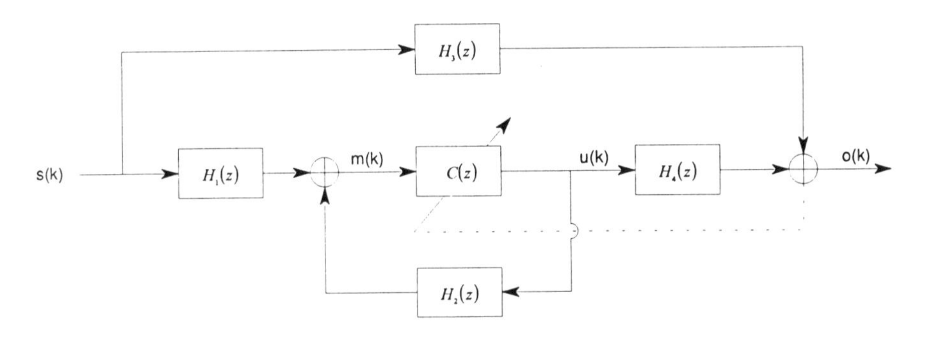

$s(k)$: primary source signal
$u(k)$: secondary source signal
$m(k)$: detector signal
$o(k)$: error signal
$C(z)$: transfer function of the controller
$H_1(z)$: transfer function of the acoustic path through primary source to detector
$H_2(z)$: transfer function of the acoustic path through secondary source to dectector
$H_3(z)$: transfer function of the acoustic path through primary source to error sensor
$H_4(z)$: transfer function of the acoustic path through secondary source to error sensor

Fig. 4.10. General Block Diagram of Adaptive Active Noise Attenuation System with Input Contamination

In general, the following assumptions must be held.

Assumptions :

(A1) $H_1(z)$, $H_2(z)$, $H_3(z)$ and $H_4(z)$ are stable rational transfer functions (by finite dimensionality) of acoustic paths.

(A2) The number of pure delays of $H_1(z)H_4(z)$ is less than or equal to that of $H_3(z)$, i.e. $\tau_1 + \tau_4 \leq \tau_3$.

The assumption (A1) is necessary and enables the acoustic systems be identified by the use of parameter identification techniques. In addition, a causal system is obtained as long as the assumption (A2) holds.

Design Procedure. A good ANC performance can be obtained if the parameters of the transfer functions of $H_1(z)$, $H_2(z)$, $H_3(z)$ and $H_4(z)$ are accurately identified so that an optimum controller $C(z)$ is derived to drive the secondary sources to compensate for the noise interference. The operational procedure of the active noise control system can be divided into two stages. The first stage is to optimize the transfer function model for the acoustic paths whereas the second stage is to optimize the transfer function model of the controller $C(z)$.

To further enhance the system's stability, the troublesome positive acoustic feedback path $H_2(z)$ from the secondary loudspeakers to the detector must be eliminated. This is possible by the use of a piezoelectric accelerometer as the primary source detection device. This sensor picks-up only the mechanical vibration rather than the actual acoustic signal, although a direct relationship between the vibration forcing signal and the acoustic sound pressure waves can be established. In this way, the accelerometer senses only the noise source vibration signal and is unaware of the acoustic signal due to secondary sources (loudspeakers). To formulate the controller $C(z)$ for noise control, the procedure is much simplified and is listed as follows:

Step 1: Estimation of the transfer function from the detector to the error sensor, $H_3(z)H_1^{-1}(z)$

Recall Eqn. 4.16 and $H_2(z) = 0$, the resultant signal at the error sensor can be expressed as

$$o(k) = H_3(k)\frac{m(k)}{H_1(z)} + H_4(z)u(k) \tag{4.17}$$

Consider an interval $0 < k \leq N_1$ while the secondary source is turned off. Here

$$u(k) = 0 \quad \text{and} \quad u(k)H_4(z) = 0 \tag{4.18}$$

Substitute Eqn. 4.18 into 4.17, the error signal is thus

$$o(k) = m(k)H_3(z)H_1^{-1}(z) \tag{4.19}$$

Consider $m(t)$ and $o(t)$ as the input and output of the transfer function, $H_3(z)H_1^{-1}(z)$, this step is equivalent to parameter identification of the unknown transfer function $H_3(z)H_1^{-1}(z)$.

Step 2: Estimation of the controller, $C(z)$ To minimize $o(k)$ as indicated in Eqn. 4.17, for the optimal noise cancellation, then

$$-m(k)H_3(z)H_1^{-1}(z) = [m(k)H_4(z)]\,C(z) \tag{4.20}$$

Similarly, the controller design for $C(z)$ can also be considered a parameter identification problem while $m(k)H_4(z)$ is available. In this case, $m(k)H_4(z)$ is obtained by the following procedure:

Consider $N_1 < k \leq N_1 + N_2$ while the reference signal $m(k)$ is transmitted through the secondary sources, that is,

$$u(k) = m(k) \tag{4.21}$$

Then, from Eqn. 4.17,

$$m(k)H_4(z) = o(k) - m(k)H_3(z)H_1^{-1}(z) \tag{4.22}$$

Since $H_3(z)H_1^{-1}(z)$ has already been obtained in the previous step, $m(k)H_4(z)$ can thus be obtained by Eqn. 4.22.

Estimation Model. Different kinds of filter models can be adopted for the ANC problem. Due to the limitation of the gradient climbing technique, the most common model applied is the FIR filter. Since a Digital Signal Processor (DSP) is used for real time ANC, the digital FIR model is normally used,

$$H(z) = \sum_{i=0}^{L-1} a_i z^{-i} \tag{4.23}$$

where L is the filter length, and a_i is the filter coefficient.

Considering that the ANC configuration is largely affected by the inherent nature of time delay, it is obvious that a large portion of the high order filter order is being using for time delay modelling. Hence, the inclusion of the time delay element in the model should prove a more appropriate approach instead of the usual high order filter.

Eqn. 4.24 proposes a general form of the modified model for the estimation of the acoustic path process [146].

$$H(z) = gz^{-d} \sum_{i=0}^{L-1} b_i z^{-ni} \tag{4.24}$$

where g is the appropriate d.c gain; d is the time delay element; L is the number of tap; and n is the tap separation.

An evaluation process for verification of this model was conducted in terms of acoustic path estimation. Results were compared for the conventional 81-tapped* FIR filter model and the modified 21-tapped† FIR filter model with $n = 4$ in Eqn. 4.24. Double tone noises of frequencies 100Hz and 250Hz were applied for this evaluation exercise. The residue error signals $o(k)$ due to the use of these filters are shown in Fig. 4.11. It is evident that the lower order modified FIR model is far better than the conventional higher order FIR filter. The time response is fast and high frequency filtering is also present, and this is considered to be an important asset for ANC. Judging from these results, the inclusion of delay and gain in the model will enhance the noise reduction performance.

---

* For conventional FIR, one tap is defined as $\left[b_i z^{-i}\right]$ where i is an integer.

† For modified FIR, one tap is defined as $\left[b_i z^{-ni}\right]$ where i and n are both integers. $n = 4$ is experimentally determined.

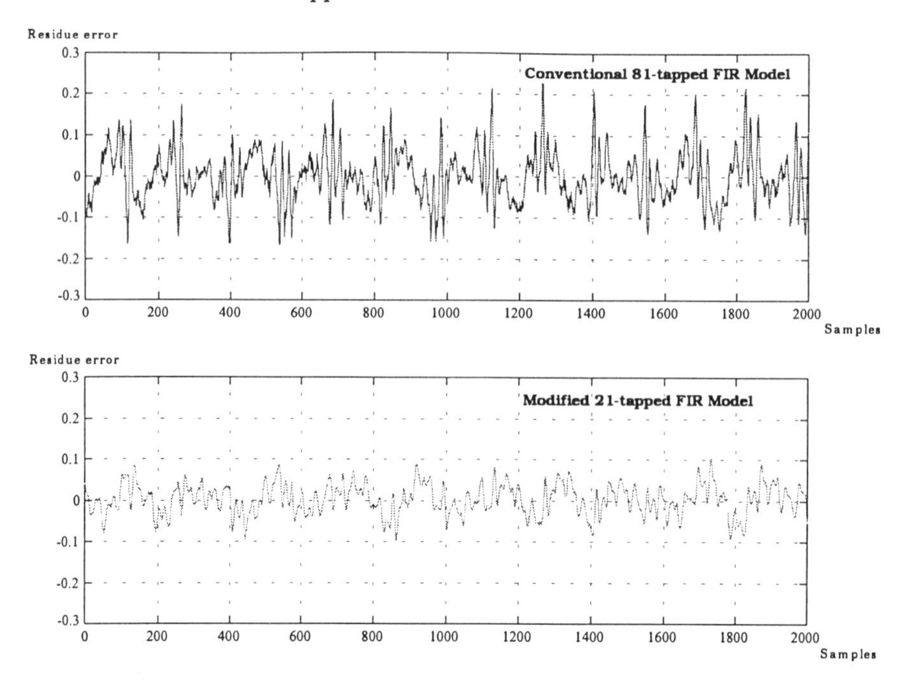

Fig. 4.11. Comparison of the Conventional and Modified FIR Models

It should be noted that the traditional gradient technique is ill suited to estimate this modified model correctly, particularly for the variables of gain and delay elements, which are initially unknown and belong to the class of multimodal error surfaces for optimization.

4.2.2 Simple Genetic Algorithm

Parameter Identification. As derived in Sect. 4.2.1, ANC is now converged into a two-stages parameter identification problem. Fig. 4.12 depicts the use of GA for such a problem.

The objective is to minimize the error between the unknown system output and the output of the estimation model. Hence, the objective function may be defined as the windowed mean square error

$$f = \frac{1}{N}\sum_{i=1}^{N}(y(k) - x(k)H(z))^2 \tag{4.25}$$

where $x(k)$ and $y(k)$ are the digitized input and output values of the unknown system, respectively;and N is the window size.

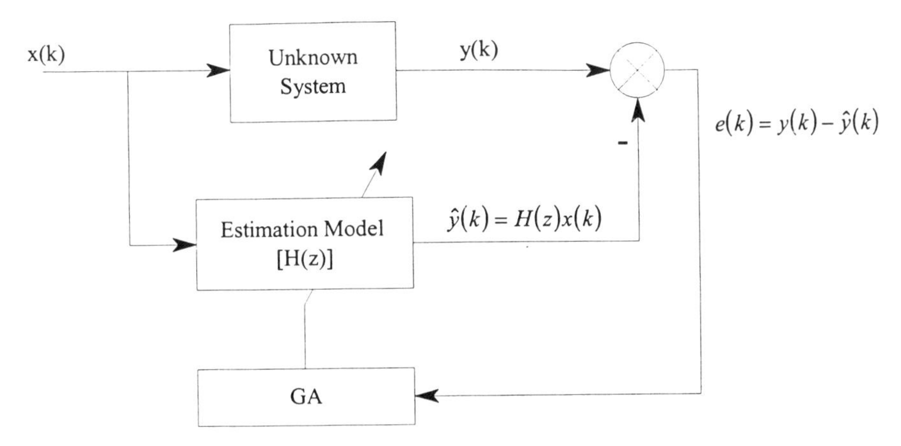

Fig. 4.12. Parameter Identification using GA

Genetic Algorithm for Parameter Identification. Based on the new model expressed in Eqn. 4.24, the structure of the chromosome is defined and formulated as follows:

$$I = \{d, g, B\} \in \Phi \subset Z \times \Re \times \Re^L \tag{4.26}$$

where $d \in [0, \alpha] \subset Z$; $g \in [0, \beta] \subset \Re$; and $B = [b_0, b_1, \ldots, b_{L-1}] \in [-1, 1]^L \subset \Re^L$ with Z and $\Re$ are the sets of integer and real numbers, respectively; α and β are the maximum range of d and g, respectively.

It should be stressed that this constrained model introduces a multimodal error surface which is considered impossible for the application of traditional gradient methods. On the other hand, GA guarantees that the objective function is globally optimized despite its multimodal nature and the constraints of the model. A brief summary of the GA operations required to achieve the optimization goal is given below:

Fitness Assignment. The fitness value of the chromosome is assigned by a linear ranking scheme [158] based on the objective function in Eqn. 4.25.

Parent Selection. The stochastic universal sampling (SUS) method [7] is applied for selection.

Crossover. The genes of the chromosome can be classified as two types: delay-gain genes $[d, g]$ and filter coefficient genes $[b_0, b_1, \ldots, b_{L-1}]$. A one point crossover operation [50] was applied to both type of genes independently.

Mutation. In natural evolution, mutation is a random process designed to introduce variations into a particular chromosome. Since the genes for delay and gain $[d, g]$ as well as those for filter coefficients are represented by integer or real numbers, random mutation [101] was applied.

Insertion Scheme. The newly-generated chromosome is reinserted into the population pool if its fitness value is better than the worst one in the population pool [50].

Termination Criterion. Since GA is a stochastic searching technique, it experiences a high variance in response time. Hence, the progress per generation is used to determine the termination of the GA.

$$t = \begin{cases} 1 & \text{if } \Gamma_k = \Gamma_{k+1} \quad \forall i \in (0, r] \\ 0 & \text{otherwise} \end{cases} \tag{4.27}$$

where Γ is the population at k-th generation.

Table 4.2. Relationship of r, Objective Value and Termination Generation

r	$(\bar{f})$	$(\bar{\tau})$
2	12.2451	161.05
4	1.3177	334.15
6	0.8662	414.60
8	0.5668	512.20
10	0.5283	579.30
12	0.5076	670.50
14	0.4894	742.90
16	0.4866	827.30
18	0.4792	933.85
20	0.4662	984.30

In order to determine a proper value of r, 20 experimental trials were conducted. The relationship between the objective mean value $(\bar{f})$, and the terminated generation mean $(\bar{\tau})$is tabulated in Table 4.2. A trade-off between the two values is made in selecting r. If the value of $r > 8$ is selected, a slight improvement in accuracy may be achieved but the real-time performance deteriorates. Hence, the empirical data indicated that $r = 8$ was a reasonable choice for terminating the GA production.

Implementation. Although the above learning procedure of the GA can achieve the parameter identification in an ANC problem, to avoid the intrinsic problem of randomness of the GA at the initial stage, and to guarantee at least some level of noise reduction at the beginning, it is necessary to combine the GA and the traditional gradient techniques in an efficient manner to achieve the ANC objective. The traditional gradient technique may not provide the required optimal performance, but its instantaneous response is an asset to the human hearing response as well as to real time control. Therefore, both GA and traditional gradient optimization procedures should be integrated for this purpose. Initially, the performance of the system using a low order FIR filter with traditional gradient optimization routines need not be optimal, and even the level of noise reduction may be low. The controller $C(z)$ will be continuously updated when a global solution is found by the GA for the modified model. This can only be realized by hardware via a communication link between these two processes. In this way, the real-time deadline will be met and an optimal noise control performance is also guaranteed. Fig. 4.13 shows the parallel hardware architecture, using two independent TMS320C30 digital signal processors for such implementation [146].

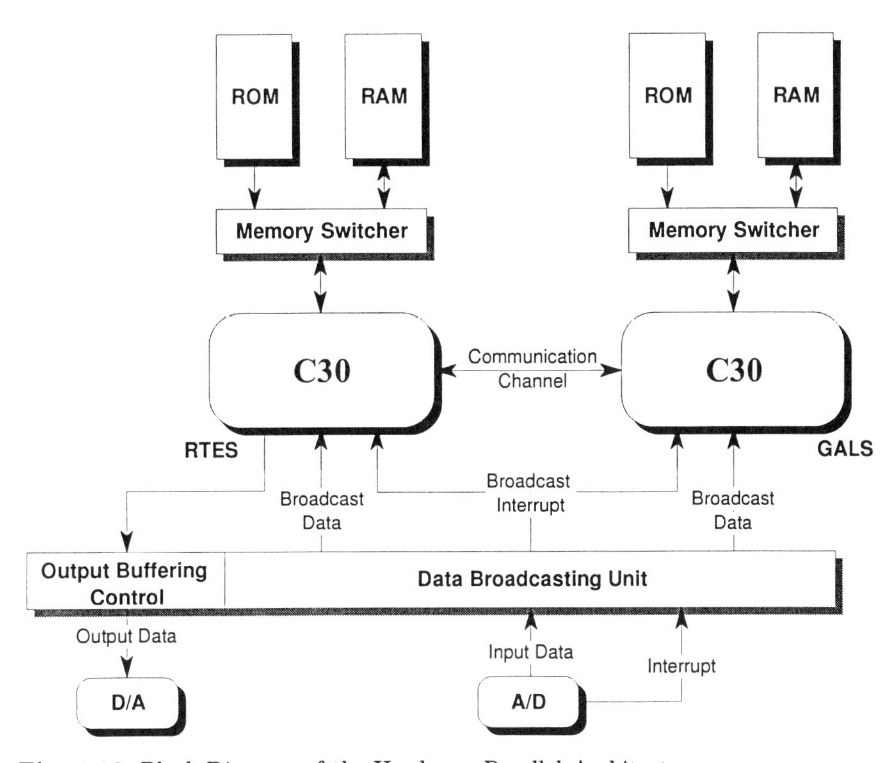

Fig. 4.13. Block Diagram of the Hardware Parallel Architecture

The architecture consists of two individual units, known as the Real-Time Executed System (RTES) and the Genetic Algorithm Learning System (GALS). RTES is used to provide a speedy, but controllable, solution of the system by conventional Recursive Least Squares (RLS) while GALS optimizes and refines the controller, in order to achieve the required optimal noise control performance. Each system is implemented using a TMS320C30 processor together with its own local memory. To prevent the data access serialization and delays that are usually experienced by each unit, a data broadcasting device was designed to handle the distribution of the external data addressed by each unit. Such a broadcasting unit releases the dependence of each processor since each one virtually has its own input device. As a result, the inter-units communication command is greatly minimized.

Statistic Monitoring Process. As the system is designed to adapt to the change of noise environment, the run-time primary source signal $m(k)$ must be monitored in order to guarantee the performance requirement. This signal is compared with the estimated signal $\hat{m}(k)$ to confirm whether there is any change of environment taking place. $\hat{m}(k)$ is calculated by

$$\hat{m}(k) = g_m \sum_{i=0}^{L-1} [\alpha_i m(k-1-ni-d_m)] \tag{4.28}$$

The parameters α_i and g_m are learned by the GA using the data sequence of $m(k)$ collected in the past optimization process. Hence, the estimated error $e(k)$ is expressed as:

$$e(k) = m(k) - g_m \sum_{i=0}^{L-1} [\alpha_i m(k-1-ni-d_m)] \tag{4.29}$$

The mean $(\bar{e})$ and variance σ^2 of the $e(k)$ within the data sequence can thus be determined. A Statistical Control procedure was established to ensure the robustness of this scheme. This assumes that the process is subject only to its natural variability and remains in a state of statistical control unless a special event occurs. If an observation exceeds the control limits, a statistically significant deviation from the normal operation is deemed to have occurred, that is when:

$$m(k) - \hat{m}(k) > \bar{e} \pm 3\sigma \tag{4.30}$$

Any change of environment will cause the restart of the RTES learning cycle automatically.

Experimental Setup and Results. The performance of the system was investigated using specifically-designed experimental equipment to realize the active noise control configuration shown in Fig. 4.14. It comprises a primary source (loudspeaker) and four additional secondary sources which were located close to the primary source with a quadpole arrangement, using four small loudspeakers. Fig. 4.14 shows the quadpole arrangement of the primary and secondary sources. The circle indicated with the mark '+' denotes the primary noise source and the other circles with the marks '-' denote the secondary sound sources [82].

The error microphone was placed perpendicular to the vertical plane of the primary and secondary sources at a distance of about 1m away from the centre of the primary source. This meant that the position of the error microphone could be in the doublet plane of symmetry in order to obtain an optimal performance [64]. The piezoelectric accelerometer was attached to

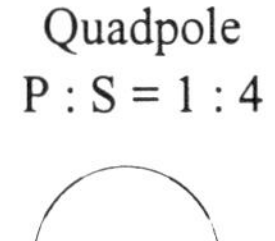

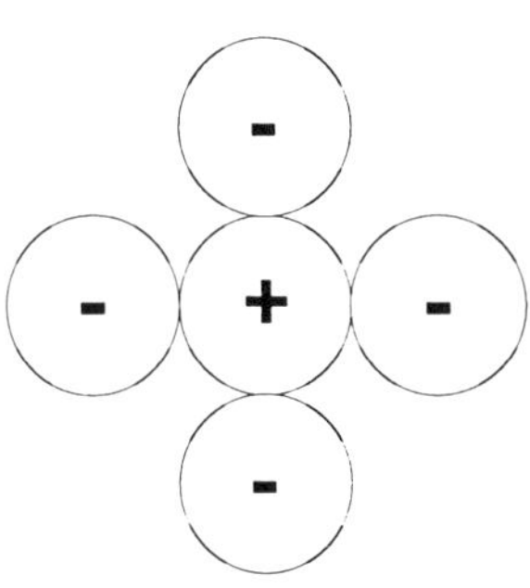

Fig. 4.14. Geometry of the Primary Sound Source 'P' (+) and the Secondary Sound Sources 'S' (-)

the primary source.

The experiments were conducted in a general laboratory with a dual tone noise signal of 100Hz and 250Hz. The sampling frequency for the Analog-to-Digital Converter was 10kHz.

The parameters of the subsystems were set as below:

- *RTES* - $H_3(z)H_1^{-1}(z)$ and $C(z)$ were modelled by traditional 21-tapped FIR filters and 1000 iterations of the Recursive Least Square algorithm were used to estimate the coefficient values.
- *GALS* - $H_3(z)H_1^{-1}(z)$ and $C(z)$ were estimated in the form of Eqn. 4.24 with a delay parameter (d) and a gain factor (g) for the modified 21-tapped FIR filter. The searching space was defined as below:

$$\begin{aligned} d &\in [0, 100] \subset Z \\ g &\in [0, 5] \subset \Re \\ B &= [b_0, b_1, \ldots, b_{20}] \in [-1, 1]^{21} \subset \Re^{21} \end{aligned}$$

The experimental noise level was recorded by a RION 1/3 Octave Band Real-Time Analyzer SA-27. Table 4.3 shows the power sum levels of all the bands and the power sum level with a frequency A-weighted characteristic when the ANC system is being turned on and off. The results are depicted in Fig. 4.15.

Table 4.3. Power Sum Value for ANC System On and Off

	A_p/dB	$A_p(\omega)/dB$
ANC – OFF	63.9	53.5
RTES – ON	61.9	47.1
GALS – ON	52.9	38.4

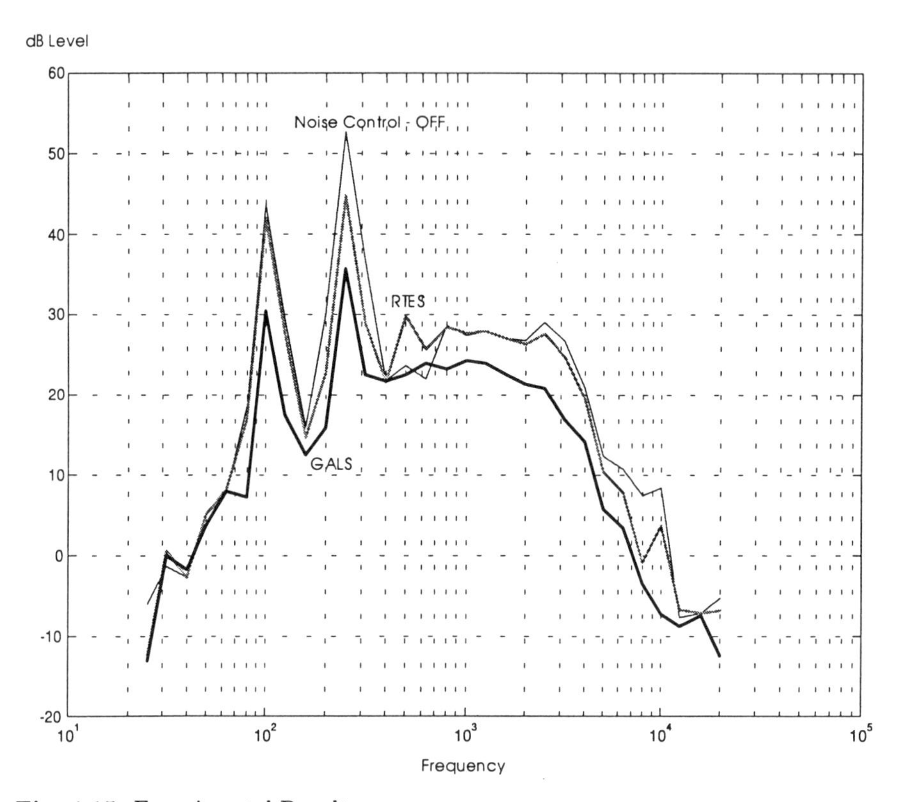

Fig. 4.15. Experimental Result

It can be seen from these results that the GA operated scheme GALS outperforms the conventional FIR filters in RTES. In addition, the dual tone frequency signals are greatly reduced by more than 15dB each. The high frequency noise is also suppressed. With the RTES scheme, this phenomenon was not observed and the general noise reduction performance was also very poor when using the equivalent low order FIR filters.

4.2.3 Multiobjective Genetic Algorithm Approach

One of the distinct advantages of the GA is its capacity to solve multiobjective functions, and yet it does not require extra effort to manipulate the GA structure in order to reach this goal. Therefore, the use of this approach for ANC makes it a very good proposition to optimize a "quiet zone" and, at the same time, alleviates the problem of selecting the error sensor (microphones) placement positions at the quiet end to achieve a good result.

Consider a multiple channel ANC system that consists of m error sensors and n secondary sources in an enclosure depicted in Fig. 4.16. The GA can be used to optimize each error sensor independently to fulfil their targets.

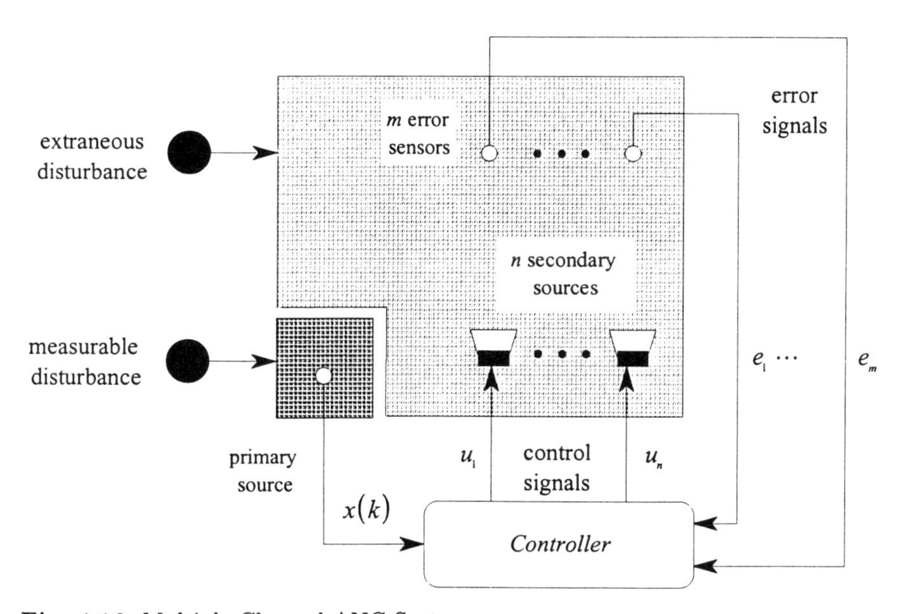

Fig. 4.16. Multiple Channel ANC System

Multi-objective Functions. At error sensor i, an objective function (f_i) is defined as below:

$$f_i = \frac{1}{N} \sum_{k=0}^{N-1} e_i^2(k) \tag{4.31}$$

where $e_i(k)$ is the acoustic signal obtained at error microphone i at time k; and N is the window size.

Instead of forming a single criterion function by lumping the objective functions with a linear or nonlinear polynomial, a multiobjective vector (F) is defined as below:

$$F = \begin{bmatrix} f_1 \\ f_2 \\ \vdots \\ f_m \end{bmatrix} \tag{4.32}$$

where f_i is defined as Eqn. 4.31, and m is the number of error sensors.

The required noise control system is applied to minimize this objective vector, i.e.

$$\min_{C(z) \in \Phi} F(I) \tag{4.33}$$

where Φ is the searching domain of the controller $C(z)$.

Genetic Active Noise control System. In order to realize Eqn. 4.33, a Genetic Active Noise Control System (GANCS) [144] which is shown in Fig. 4.17 is proposed. GANCS consists of four fundamental units, namely, Acoustic Path Estimation Process (APEP), Genetic Control Design Process (GCDP), Statistic Monitoring Process (SPM) and Decision-Maker (DM).

The design concept of the controller is basically composed of two processes:

1. Acoustic Path Estimation Process for acoustic paths modelling which has the same procedure as Sect. 4.2.1.
2. Genetic Control Design Process for controller design which has a similar structure to that described in Sect. 4.2.2. The main difference in the fitness assignment is explained in more detail in the following sections.

The idea is to model the acoustic paths using FIR filters while the development of the controller relies on this obtained modelling result. The Statistic Monitoring Process monitors the change of environment to ensure the system's robustness as explained in Sect. 4.2.2. The Decision Maker provides an interface so that the problem goal can be defined to fine-tune the optimization process of the GCDP.

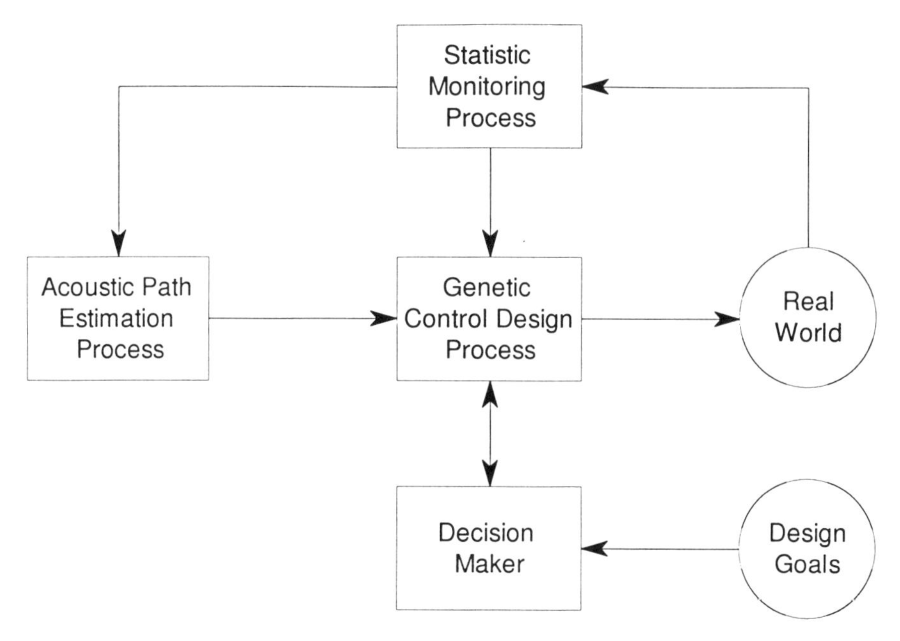

Fig. 4.17. Block Diagram of GANCS

Decision Maker. Considering that the Multi-channel Active Noise Control is a multiple objective problem, a Pareto-solution set would be obtained. The Decision Maker selects the solution from this set of non-dominate solutions. In general, the goal attainment method can be applied to achieve the best global noise suppression result. Consider that the goals of the design objective are expressed as

$$V = (v_1, v_2, \dots, v_m) \tag{4.34}$$

where v_i is the goal for the design objective f_i.

The non-dominate solutions are compared with the λ values which are expressed below:

$$f_i - \lambda w_i \leq v_i \tag{4.35}$$

where w_i is weighting factor for v_i.

This is the targeted residue noise power level at error sensor i to be reached. Due to different practical requirement, a specific goal should be set. The system will be tuned to meet such goal requirement. This can be illustrated by a simple example:

$$\begin{aligned} w_i &= 1 \quad \forall i \in [1, m] \subset Z \\ v_i &= \alpha \end{aligned}$$

$$v_j = 2\alpha$$

which means a higher requirement of silence is assigned in error sensor i.

GCDP – Multiobjective Fitness Assignment. The structure of the GCDP is similar to that of the GA applied in the previous section. The only difference is that a multiobjective rank-based fitness assignment method, as explained in Sect. 3.2, is applied. With the modified ranking method, the non-dominate solution set can be acquired.

Experiment Results. Experiments were carried out to justify the multiobjective approach. A single tone noise signal of 160Hz was used to demonstrate its effectiveness. The error microphones, P1 and P2 were placed 1.3m and 0.5m above ground, respectively. The overall geographic position of the set-up is shown in Fig. 4.18.

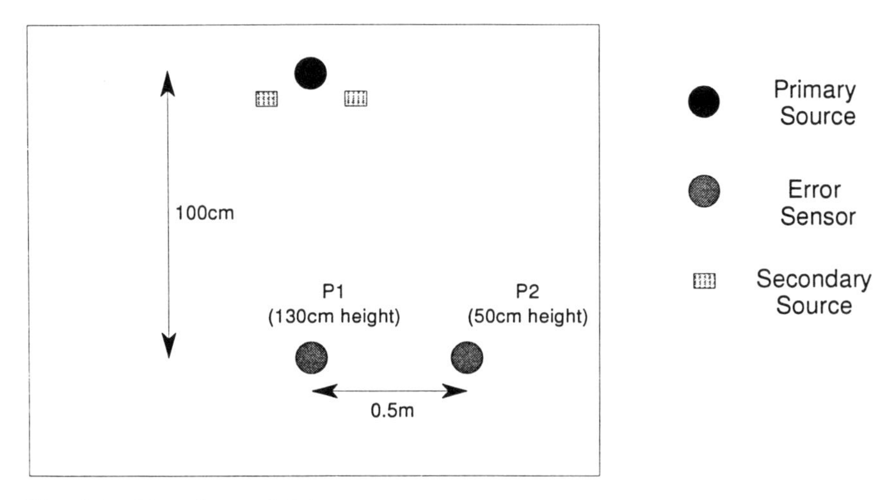

Fig. 4.18. Experimental Set-up

The purpose of this particular set-up was to justify the multiobjective approach in ANC, therefore, three tests were conducted:

1. to optimize P1 location only;
2. to optimize P2 location only; and
3. to optimize P1 and P2 simultaneously using the multiobjective approach

Figs. 4.19 and 4.20 depict the noise reduction spectra of P1 and P2 for case (1) and case (2), respectively. Fig. 4.21 shows the compromising effect when both P1 and P2 are optimized by the multiobjective approach. From Table 4.4, it can be observed that the reduction levels on P1 and P2 are under consideration for a simultaneous optimization procedure when the multiobjective approach is applied.

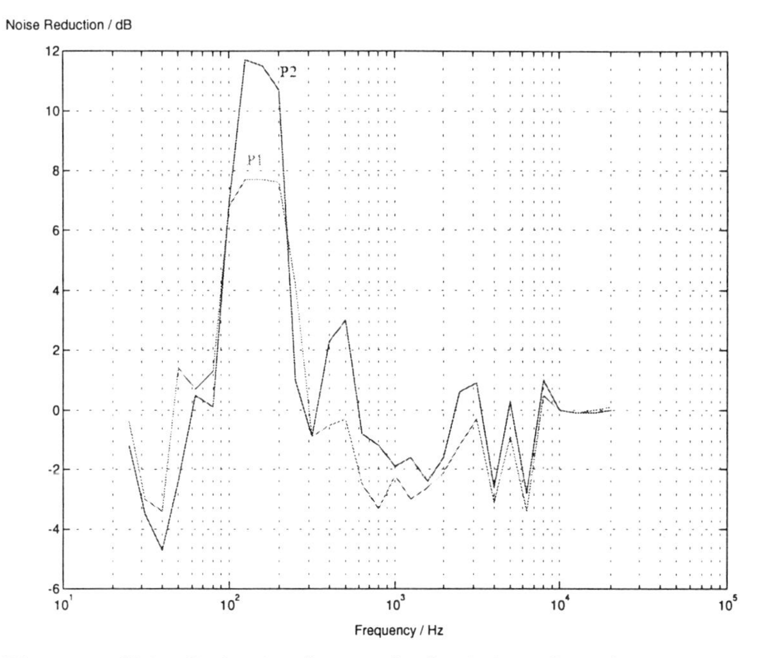

Fig. 4.19. Noise Reduction Spectra for Optimizing P1 only

Table 4.4. Noise Reduction for Different Cases

	Position 1		**Position 2**	
	A_p/dB	$A_p(\omega)/dB$	A_p/dB	$A_p(\omega)/dB$
Noise without Control	92.0	78.8	87.7	74.5
Optimization on P1	84.4 (-7.6)	71.2 (-7.6)	76.7 (-11.0)	63.2 (-11.3)
Optimization on P2	90.5 (-1.5)	77.2 (-1.6)	67.5 (-20.2)	54.9 (-19.6)
MO Approach	87.0 (-5.0)	73.9 (-4.9)	72.7 (-15.0)	59.2 (-15.3)

To further demonstrate the effectiveness of the multiobjective approach for ANC, the previous experimental set-up was altered so that the new geographical sensor placements were arranged as shown in Fig. 4.22. The particular layout was purposefully arranged in a different form from the one above. The idea was to test the functionality of the multiobjective approach such that the wavefront in this case would be substantially different since the microphone P1 was placed 0.8m ahead of P2. In this way, the noise reduction result would have to be compromised in a more weighted manner than the

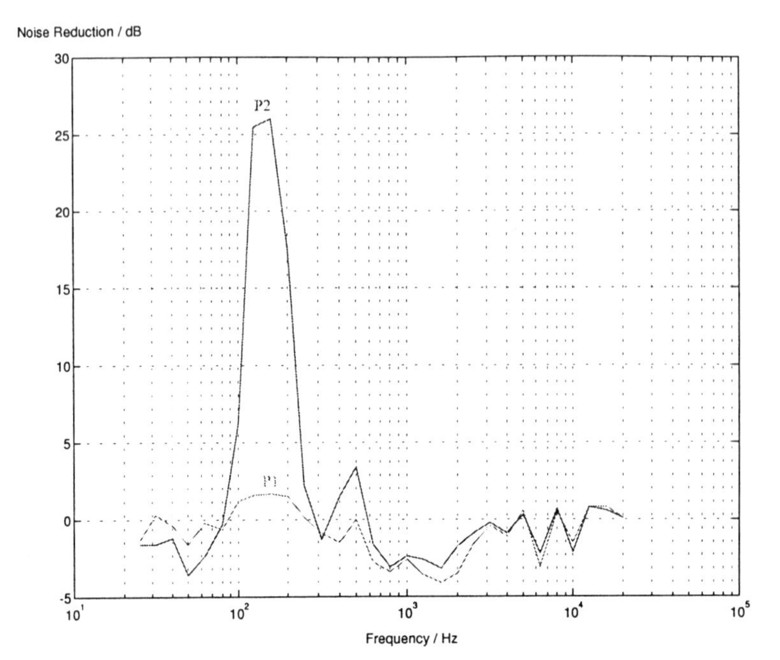

Fig. 4.20. Noise Reduction Spectra for Optimizing P2 only

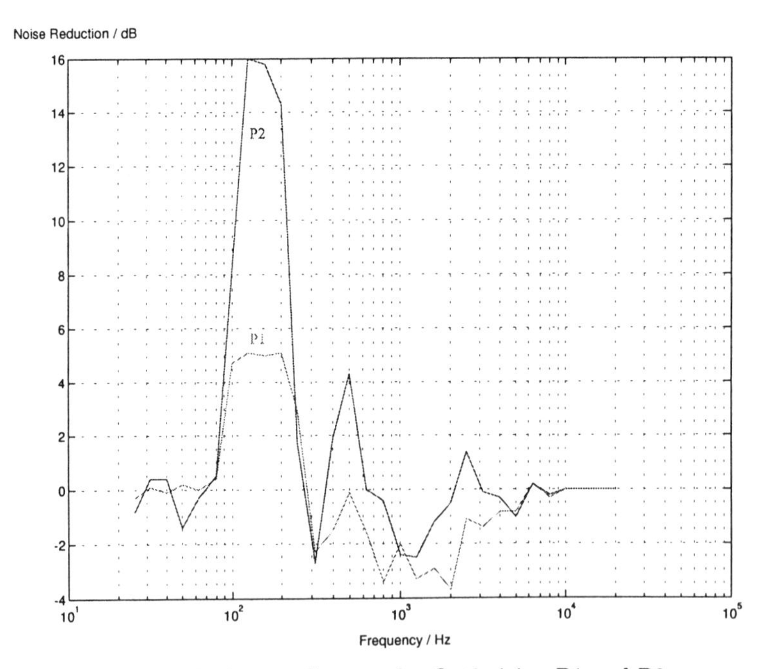

Fig. 4.21. Noise Reduction Spectra for Optimizing P1 and P2

previous case.

Fig. 4.23 shows the result of noise reduction for optimizing the location at P1 only and Fig. 4.24 depicts the equivalent result at P2. A typical result obtained (candidate 4 in Table 4.5) from the multiobjective approach is shown Fig. 4.25. The overall noise reduction results are tabulated in Table 4.5.

It can be seen from the results obtained in this case that they are quite different when compared to the previous case. There was some noise reduction at P2 while P1 was optimized, but the situation was not reversed at P2 for P1. From Table 4.5, it was a slander for P1 as there was a slight increase of noise level at that point.

However, much more compensated results were obtained from the multiobjective approach. There were five possible solutions which were yielded in the population pool by the effort of the GA. These are also tabulated in Table 4.5. It can be seen that candidate 4 would be an evenly balanced result for P1 and P2 as there is only 1–2 dB difference between them. On the other hand, candidate 1 is considered to be at the other extreme. While P2 reached a 20dB reduction, P1 was considered inactive in the case. The result obtained from candidate 3 is considered to be similar to candidate 4. The candidates 2 and 5 possess the results of a mirror image of each other at P1 and P2. The overall results of the multiobjective approach can be summarized by the trade-off curve as shown in Fig. 4.26. It can be concluded that all five candidates are practically applicable, but the final selection has to be decided by the designer via the decision maker so that the best selected candidate is used to suit that particular environment.

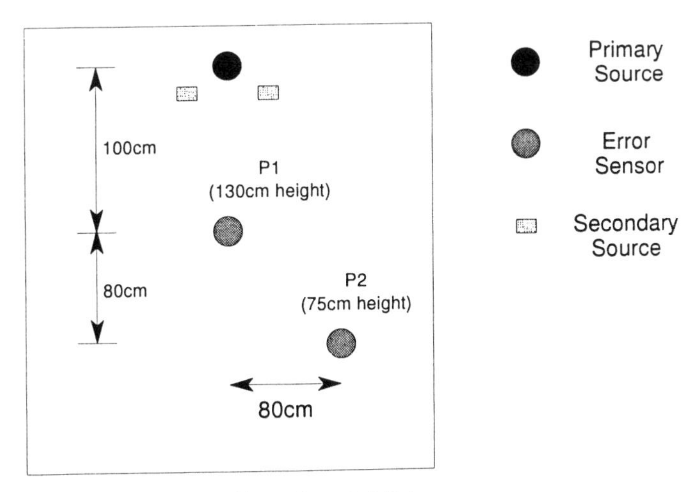

Fig. 4.22. Another Experimental Set-up

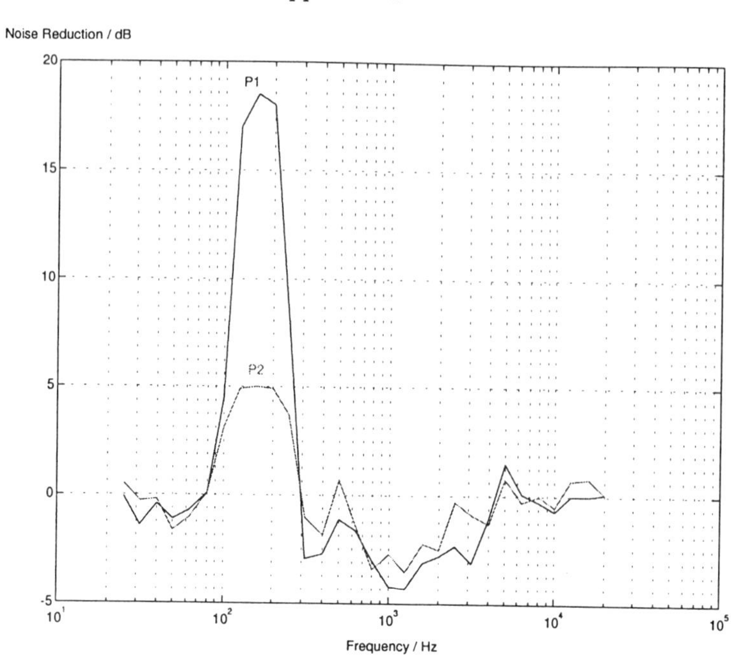

Fig. 4.23. Noise Reduction Spectra for Optimizing P1 only

Table 4.5. Noise Reduction for Different Cases

	Position 1		Position 2	
	A_p/dB	$A_p(\omega)/dB$	A_p/dB	$A_p(\omega)/dB$
Noise without Control	91.3	78.0	92.6	79.3
Optimization on P1	73.9 (-17.4)	61.8 (-16.2)	87.7 (-4.9)	74.4 (-4.9)
Optimization on P2	92.9 (+1.6)	79.6 (+1.6)	71.3 (-21.3)	58.1 (-21.2)
MO Candidate 1	91.4 (+0.1)	78.1 (+0.1)	72.0 (-20.6)	58.8 (-20.5)
MO Candidate 2	88.7 (-2.6)	75.4 (-2.6)	79.9 (-12.7)	66.6 (-12.7)
MO Candidate 3	85.7 (-5.6)	72.6 (-5.4)	82.0 (-10.6)	68.6 (-10.7)
MO Candidate 4	84.2 (-7.1)	71.1 (-6.9)	83.9 (-8.7)	70.6 (-8.7)
MO Candidate 5	73.8 (-17.5)	61.9 (-16.1)	87.1 (-5.5)	73.8 (-5.5)

4.2.4 Parallel Genetic Algorithm Approach

While we have tackled the multiobjective issue of the GA in an ANC, our effort is now turned to the computational problem of the GA in an ANC. One of the problems that it is usually encountered with sequential computation of a GA is that it is generally recognized as a slow process. Such a deficiency can

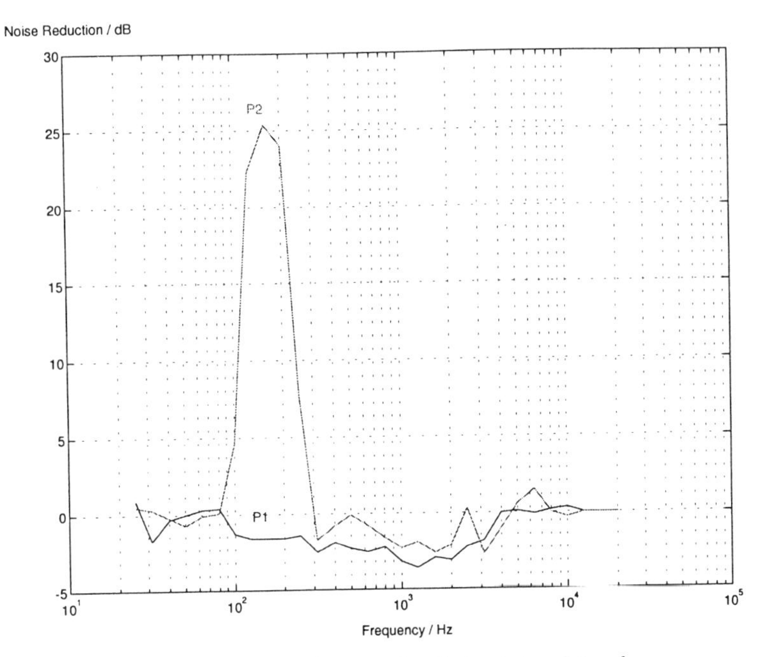

Fig. 4.24. Noise Reduction Spectra for Optimizing P2 only

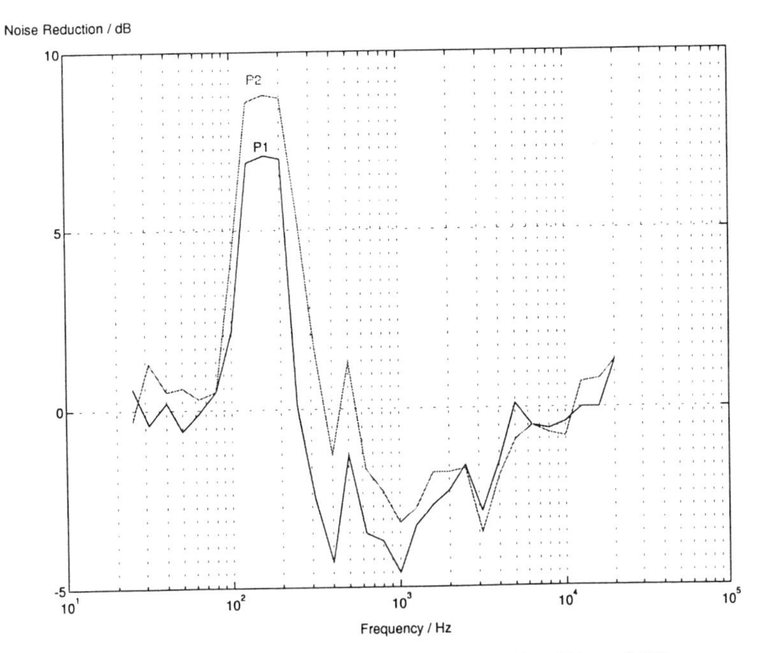

Fig. 4.25. Noise Reduction Spectra for Optimizing P1 and P2

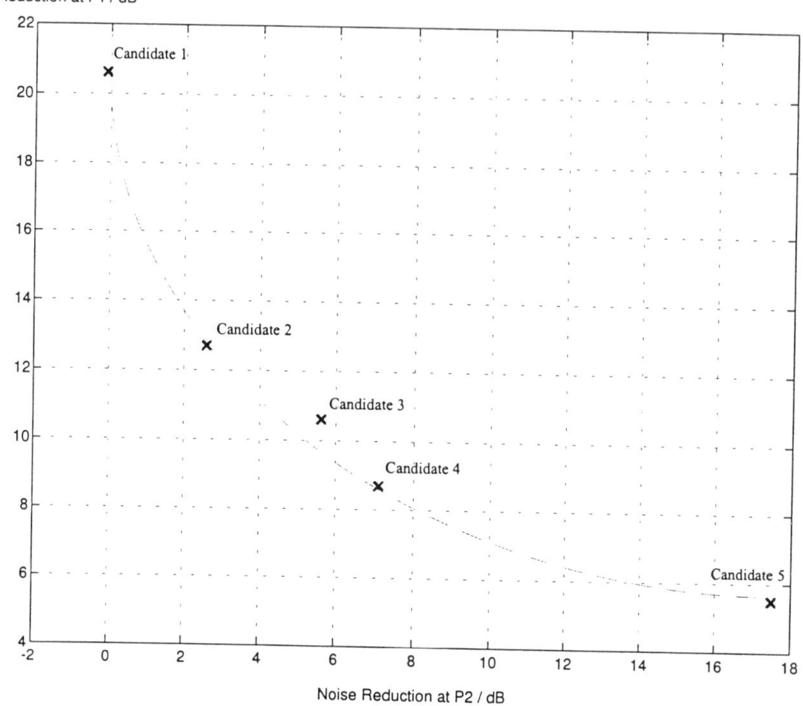

Fig. 4.26. Noise Reduction Trade-off between P1 and P2 due to Multiobjective Approach

be readily rectified by implementing parallel computing architecture. Considering that the GA itself already possesses intrinsic parallelism characteristics, it requires no extra effort to construct such parallel architecture to gain the required speed for practical uses. There are a number of GA-based parallel methods that can be used to enhance computational speed (as explained in Sect. 3.1). The global parallel GA (GPGA) is recommended for this application, in the sense that a multiple of system response is guaranteed.

The problem of a slow GA computation is not really caused by genetic operations such as selection, recombination, mutation, fitness assignments and so on. In fact, any one of these operations requires little computational time. The main contributor to the time consumption is the actual numeric calculation of the objective functions. This is particularly apparent when a number of objective functions have to be simultaneously optimized such that their functions may even be nonlinear, constrained and discontinuous.

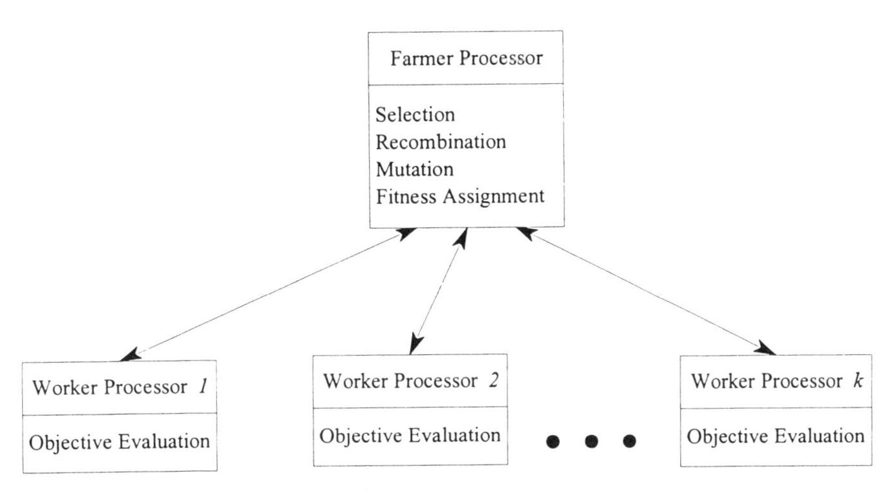

Fig. 4.27. Farmer-and-Workers Structure

Fig. 4.27 shows the farmer-and-worker structure for such global parallelism architecture which is well suited for this application.

The GPGA treats the entire population as a single breeding mechanism. The purpose of the farmer-processor (FP), being a master control unit, is to generate the new chromosomes while the worker-processors (WP) are used to evaluate the objective values of the chromosomes. The objective values of the new chromosomes are then returned to the FP for the process of reinsertion and the necessary fitness assignment. In this way, the time-demanding process of objective evaluation is now handled by a dedicated WP, and hence improves the computational speed in the order of a multiple fashion if a number of WPs are used.

Parallel Hardware Approach. Based on the farmer-and-worker parallel structure, a scalable multiple digital signal processing (DSP) based parallel hardware architecture has been developed [68]. The overall parallel architecture is shown in Fig. 4.28.

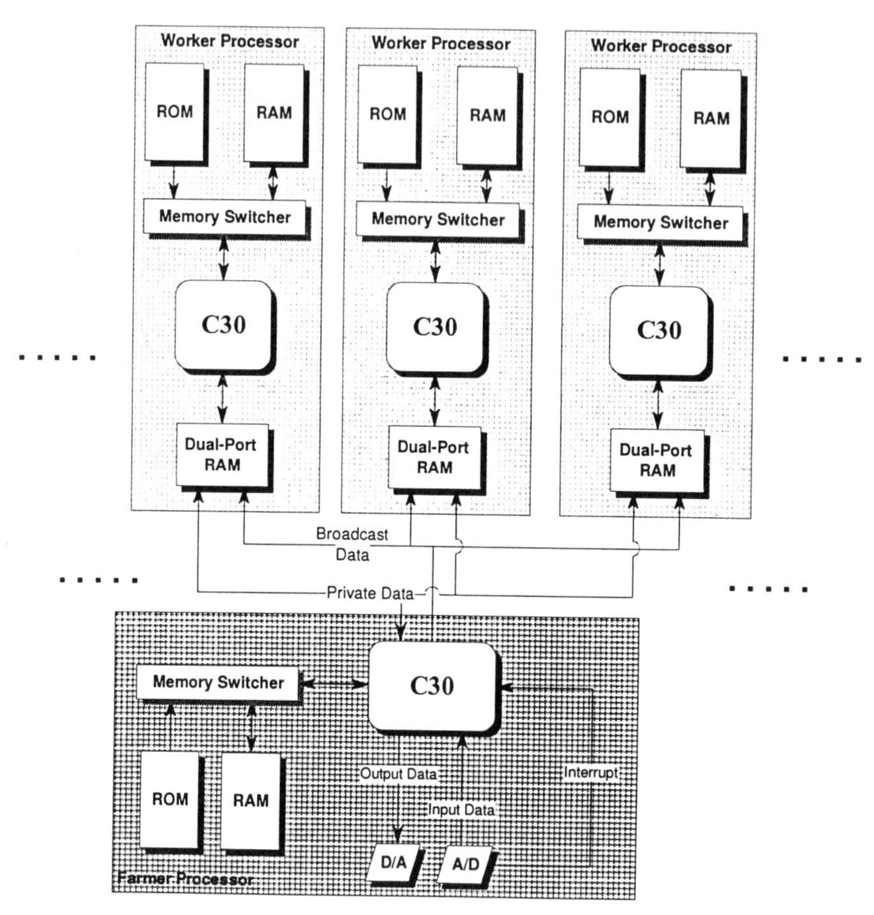

Fig. 4.28. Active Noise Control System

This architecture was formed by the use of Texas Instruments' TMS320C30 (C30) DSP processors. A C30 processor was designed as a FP unit and this handles all the input/output (IO) activities, as well as generating the new chromosomes. The FP also co-ordinates the data flow between the real-world and the system. By the use of the C30's two independent external bus interface ports: Primary Bus and Expansion Bus, the external IO address, local memory address and WP address are separated. The operations of the two buses are basically identical except that the Primary Bus address space that is 24-bit in width and the Expansion Bus address space that is 13-bit in width. The local memory and all the WPs can be accessed through the Primary Bus. However, all the IO peripherals are interfaced by the Expansion Bus.

The objective values of the chromosomes are evaluated by the C30s as WPs. These values return to the FP for the process of reinsertion and fitness assignment. Each WP has its own local memory units and associated dual-port RAM. The reason for using this device is to reduce multi-DSP interfacing complexity and to provide a scalable feature so that the inter-processor communication bandwidth can thus be increased. The other advantage of using a dual-port RAM is that it ensures the FP can communicate with the WP in the conventional memory access operating manner.

In this configuration, the main action of the FP is only to read/write data on the different memory segments so that the information can be easily accessed by a number of WPs. In our design, the memory size of the dual-port RAM is 2K words and the address Bus width of the FP is 24 bits, which implies that 16M bytes memory can be accessed through the Primary Bus. Therefore, the maximum number of WPs that can be handled by the FP is 8192 (16M/2K)!

As the number of WPs increases, so the interprocessor communication overhead increases substantially. In order to reduce the communication overhead, two types of data are classified: broadcast data and private data. Each type of data uses its own addressing method. The broadcast data are used typically during the initialization stage when all the initial data and parameters are fed into WPs. Therefore, as the number of WPs increases, the initialization processing time remains constant. The private data type is used by the FP to update the individual WP's optimized result.

This GA hardware platform using DSP chips should serve as an ideal test-bed facility for system design and development before dedicated system-specific hardware is built.

Filter Model. Another filter model, IIR filter, was tried in this experiment. Due to the multimodality of the IIR filter error surface, the GA is well suited to optimize the filter coefficients to search for global optima [107, 157]. An IIR filter can be constructed in lattice form and this structure is shown in Fig. 4.29.

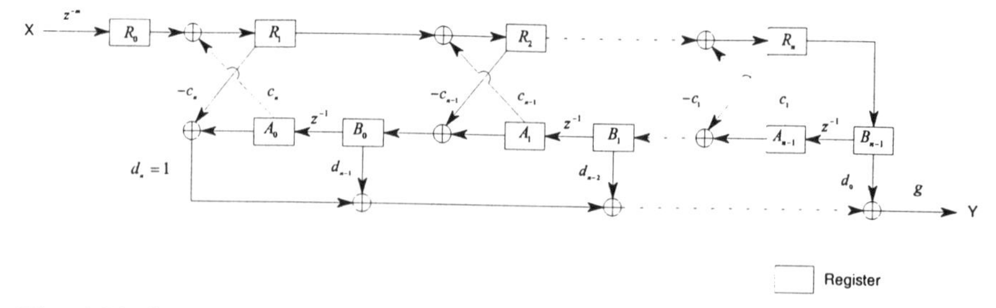

Fig. 4.29. Lattice Form of IIR Filter

Similarly, the delay and gain elements are already embedded in this filter model

$$H(z) = gz^{-m}\frac{B(z)}{A(z)} \tag{4.36}$$

$$= gz^{-m}\frac{\sum_{i=0}^{n} b_i z^{-i}}{\sum_{j=0}^{n} a_j z^{-j}} \tag{4.37}$$

where m is an additional delay; g is the gain; a_i, b_i are the IIR coefficients ($a_0 = b_0 = 1$), and c_i, d_i are the lattice filter coefficients which can be obtained by

$$c_i = a_i^{(i)}$$

$$d_i = b_i + \sum_{j=i+1}^{n} d_j a_{j-i}^{(j)}$$

with $a_i^{(n)} = a_i$, $\forall i = 1, 2, \ldots n$ and the recursive function

$$a_j^{(i-1)} = \frac{a_j^{(i)} + c_i a_{i-j}^{(i)}}{1 - c_i^2} \tag{4.38}$$

Referring to Fig. 4.29, for a n-poles and n-zeros system, we can obtain the following equations.

$$R_0 = x(k-m) \tag{4.39}$$

$$R_i = R_{i-1} + C_i A_i \qquad i = 1, \ldots, n \tag{4.40}$$

$$B_i = -C_{i+1} R_{i+1} + A_{i+1} \qquad i = 1, \ldots, n-1 \tag{4.41}$$

$$B_n = R_n \tag{4.42}$$

The output of this lattice IIR filter is computed by Eqn. 4.43.

$$y(k) = g\left\{-c_1 R_1 + A_1 + \sum_{i=1}^{n} d_i B_i\right\} \tag{4.43}$$

The principle advantage of lattice filters over alternative realizations such as Direct-Form, parallel and cascade, is that stability can be maintained simply by restricting the lattice reflection coefficients to lie within the range +1 to -1. In addition, the lattice form is known to be less sensitive to coefficient round-off. There are totally $(2n+2)$ parameters (m, g), $(c_1, c_2, \ldots, c_n)$, $(d_0, d_1, \ldots, d_{n-1})$ to optimize for a n-th order IIR filter.

Chromosome Representation. This is used to optimize the IIR parameters using a GA. These parameters can be coded into genes of the chromosome for GA optimization. There are two different data types. The unsigned integer delay gene, $[m]$, is represented by a 16-bit binary string and each parameter in the real-valued coefficient genes $[g, c_i, d_i]$ is represented by a 32-bit binary string. The parameters are constrained as

$$\begin{aligned} m &\in [0, m_{max}] \subset Z^+ \\ |g| &\in [g_{min}, g_{max}] \subset \Re^+ \\ C &= [c_1, c_2, \ldots, c_n] \in (-1,1)^n \subset \Re^n \\ D &= [d_1, d_2, \ldots, d_n] \subset \Re^n \end{aligned} \tag{4.44}$$

where $m_{max} = 100$, $g_{min} = 0.01$, and $g_{max} = 2.0$.

The structure of the chromosome (I) is thus formulated as follows:

$$I = \{m, g, C, D\} \in \Phi \subset Z \times \Re \times \Re^n \times \Re^n \tag{4.45}$$

where Φ is the searching domain of the controller $C(z)$.

For this application, 4th-order IIR filters were used to model the acoustic path and the controller. Hence, $n = 4$. With similar operations, the GA is ready to go.

Experimental Results. This formulation was used for the investigation. The ANC noise reduction result due to this approach is shown in Fig. 4.30. It can be demonstrated that the parallelism of the system could greatly reduce the computation time with a number of worker processors.

4.2.5 Hardware GA Processor

Thus far, the study of the GA in an ANC application has been successfully carried out. The results are very encouraging for developing an industrial/commercial product. However, the cost for realizing the goal is currently quite expensive, as the computers or signal processors etc. that have been incurred in the study would not be economically viable. Furthermore, for an industrial product, the ANC system must be implemented in a VLSI chip version in order to be marketable. Therefore, some means of cost cutting exercise must be sought in order to translate this technology into practice.

However, the development cost of VLSI chip does not present less cost in a dollar sense. The number of iterations of development must be limited to a

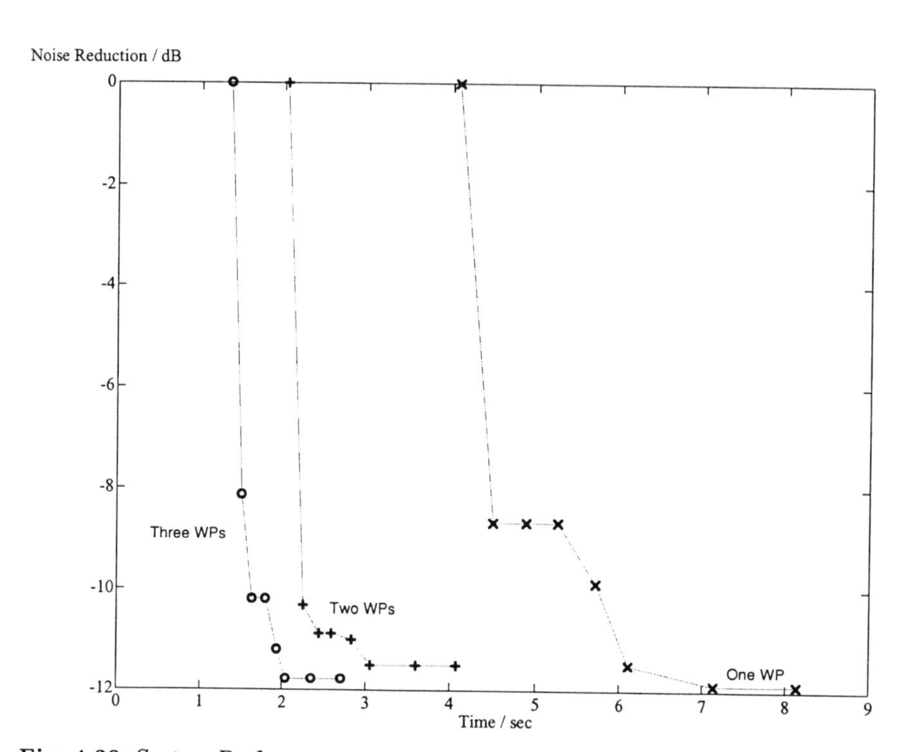

Fig. 4.30. System Performance

minimum to avoid budgeting over expense. Therefore, the development of a low cost version by the use of Field Programmable Gate Arrays (FPGA) could be ideally applied. The version of development can be realized based on the methodology that has already been described. This work is now being actively carried out, and the implementation outlined here illustrates its feasibility for the development of the future ANC systems.

It has become clear that the GA provides a means to evolve solutions to search and optimize problems instead of using other fixed mathematically derived algorithmic procedures. It is also well understood that the GA possesses a number of computational dependable schemes in order to fulfil its goal. To further improve the performance of the GA in terms of computation time, a hardware GA processor is therefore a clear solution. One of the best ways to implement a GA in hardware is to construct a model using the hardware descriptive language VHDL and to synthesize the model in FPGA technology. The realization of Global GA, see Fig. 3.1, and Migration GA, see Figs. 3.2–3.4, can be easily achieved. In this particular design, the Actel FPGA [1] using Synopsys [121] compiler has been adopted. Fig. 4.31 shows the basic design flow of synthesizing an VHDL design description into an FPGA using the Synopsys Design Compiler, FPGA Compiler, or Design Analyzer.

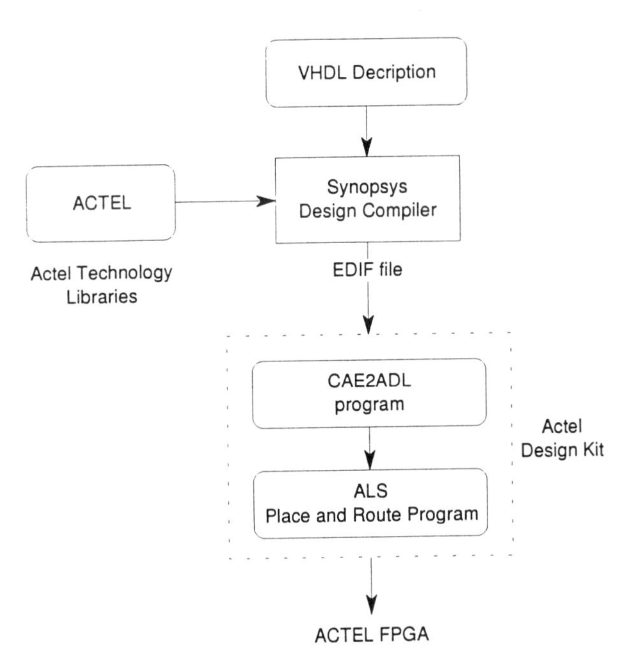

Fig. 4.31. Actel/Synopsys Design Flow

This design flow is based on the Parallel Global GA that is described in Sect. 4.2.4. This architecture is particularly useful for Single-Instruction-Multiple-Data (SIMD) design in which all the offspring are able to proceed with the same fitness evaluation process. The overall hardware for the Hardware GA processor is shown in Fig. 4.32. This is a flexible modular structure that can be further expanded if necessary. This architecture consists of three different modules, namely Fitness Evaluator (FE), Objective Function Sequencer (OFS) and Genetic Operators (GO). Each of the modules is implemented with an FPGA. This modular structure can easily lead to varied application environments by replacing a suitable module, and yet, the demand of computation power is met.

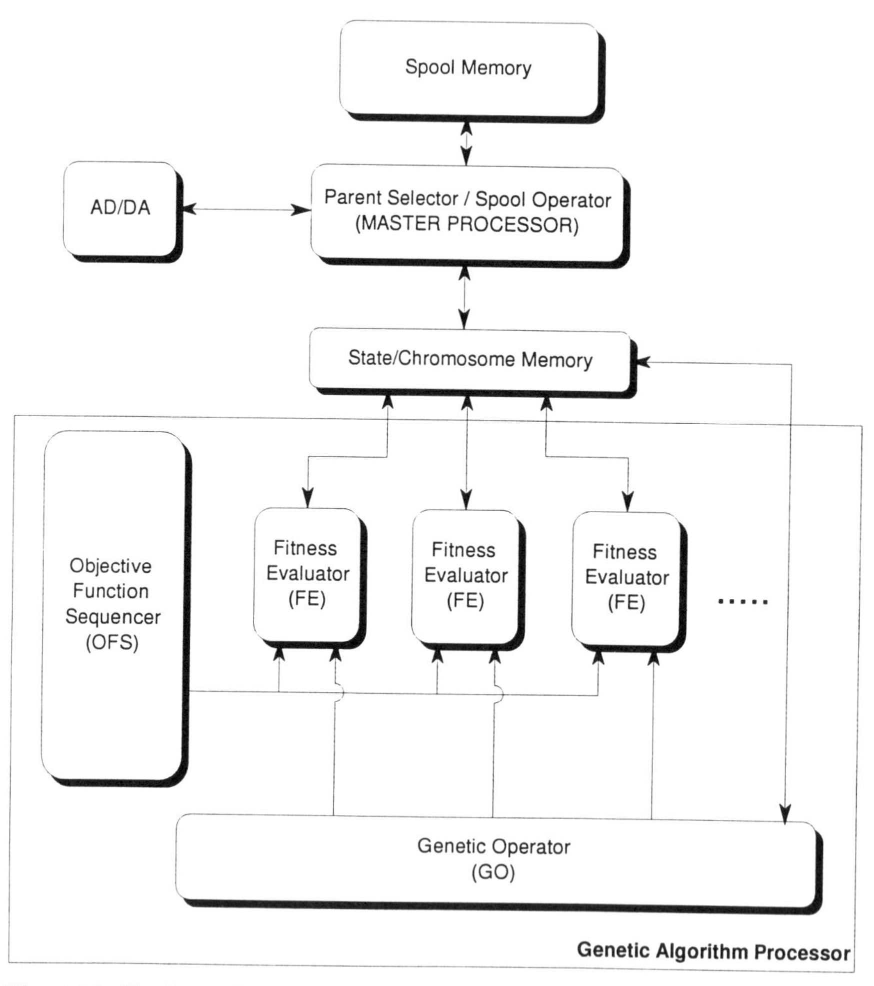

Fig. 4.32. Hardware Global GA

The relationship between the operation and inter-module of the GAP can be described as follows: In each GA cycle, the parents are selected by the external processing unit (parent selector). These parents are then fed to the GO module. A simplified structure diagram for the GO is shown in Fig. 4.33. Two GA operations, namely the mutation and crossover, are processed by the GO while a uniform crossover is adopted to exchange the genetic information of the parents. The action of mutation is then applied for altering the genes randomly with a small probabilistic value. Knowing that one difficulty of the GO would be the hardware implementation of a realistic Guassian distributed random number generator in the FPGA. A linear distributed function is therefore proposed for such a design. The details of this implementation are described in Appendix C.

Once the crossover and mutation operations have been completed in the GO, the required offspring are born. The quality of these offspring is then examined by the FE via a fitness evaluation process. This is usually a cost function in the least square form. In this case, the hardware construction involves only the development of adders and multipliers which can be easily constructed by the use of FPGA chips. The actual sequence of calculating the fitness values is specified by the instructions of a sequencer design of the OFS.

The architecture of the OFS can also be built using the FPGA chip. The OFS provides the sequence to perform the fitness evaluation of the FE. This evaluation is only completed when the fitness value of each offspring has been calculated. At this stage, all the offspring, together with their fitness values are then fed into the external state/chromosome memory for insertion into the population pool.

This loosely coupled module structure enables the GAP design to be easily fitted into various application environments in order to suit different types of computation power requirements. Since this architecture has been purposely designed not only for ANC, but also for general engineering use, the design of the OFS sequencer can be changed according to the style of the fitness functions, which in itself is a problem dependent element.

The advantage of this set up is that the units of the FE and the GO remain unchanged, despite the fact that they may be used for other applications. Furthermore, it is a scalable architecture which provides enhanced versatility for the GAP to be designed to tackle complex real-time applications. As for the time-demand problem, more FEs can be added in order to fulfil the computation requirement.

For the purposes of Active Noise Control (ANC), the lattice filter model is adopted, and the formulation of the chromosome is based on the Sect. 4.2.4.

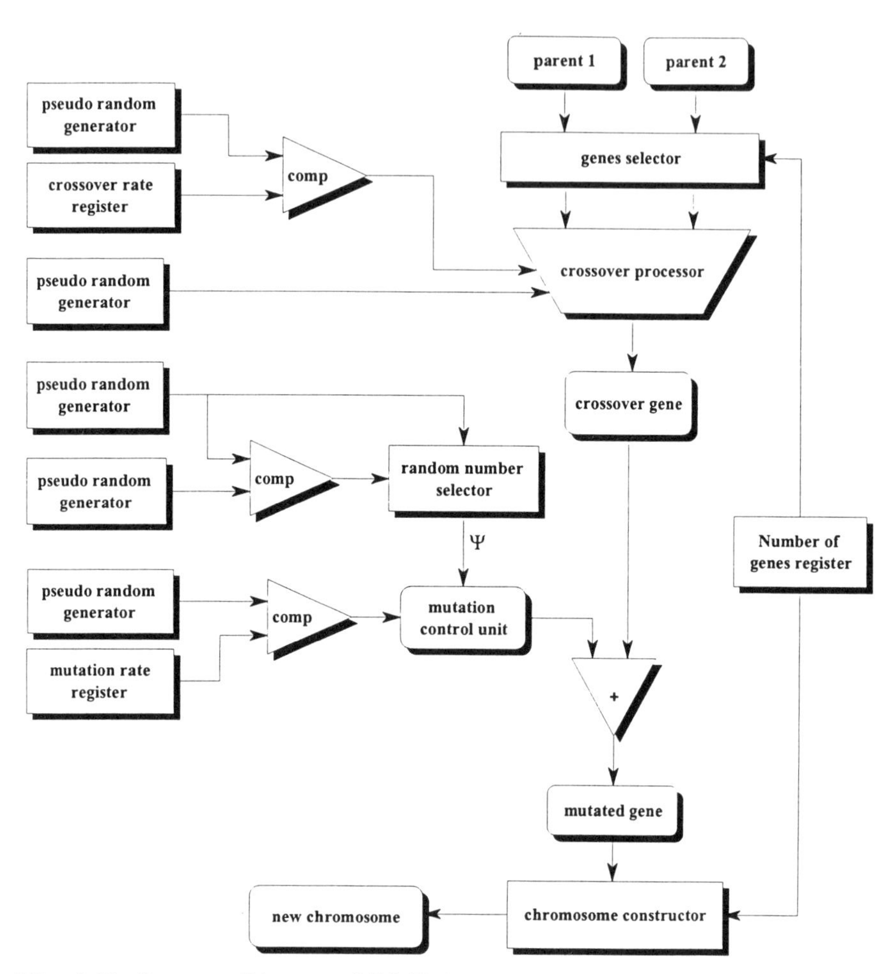

Fig. 4.33. Structure Diagram of GO Unit

Fitness Function. The fitness function reflects the likeness between the output of the estimated model and the real output. In mathematical terms, it is expressed as:

$$f = \sum_{i=1}^{N} (y_i - \hat{y}_i)^2 \tag{4.46}$$

where $\hat{y}_i$ is the estimated output from a filter model; y_i is the real output; and N is the sample window. In our experiments, $N = 400$.

Fitness Evaluator. To realize Eqn. 4.46, the design of an FE involves the design of a multiplier and an adder. A development of a high speed multiplier is the major effort for improving the computing speed. Hence, it is the main core of the FE design. In our design, a high pipelined multiplier is proposed as indicated in Fig. 4.34. The multiplier algorithm is based on the redundant binary representation [140], which is described in Appendix D. To compensate for the propagation delay of the logic modules in the FPGA, a three-stage pipeline is applied for the design to improve the performance speed.

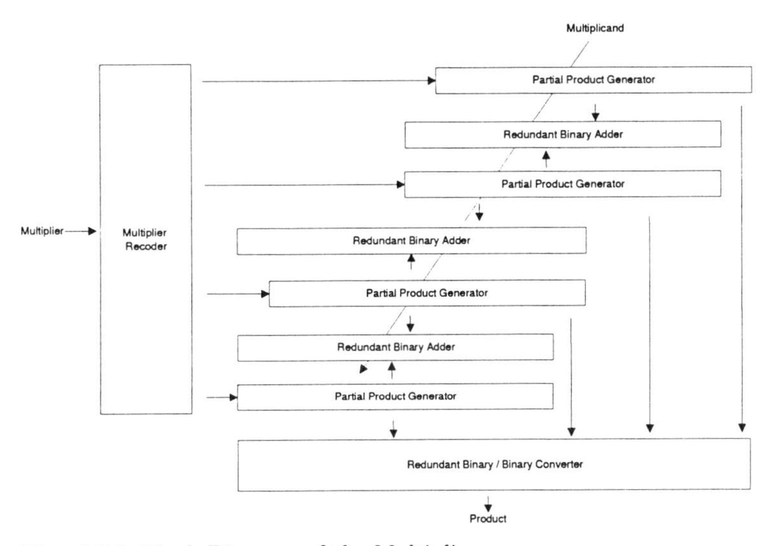

Fig. 4.34. Block Diagram of the Multiplier

Objective Function Sequencer. Based on the pipeline structure, a good sequencer design means a high efficiency of multiplier pipeline usage. As a consequence, this improves the GAP's overall performance. Therefore, it was necessary to compromise between the system performance (filter order) and the computation demand (number of steps required for completing the evaluation).

In our design, a 4x4 IIR lattice filter based on Eqn. 4.43 was applied. A computation step table, Table 4.6, was formed. The number of steps needed to complete one cycle of the fitness function evaluation are stated in this table. There are 15 computation steps in total. Since a three-stage pipeline multiplier is designed in the FE, the executing sequences are stated in Table 4.7 which was developed on the basis of Table 4.6.

Table 4.6. The Computation Step Table of the 4X4 IIR Lattice Filter

state0	read x
state1	tmp1=x+c1a1
state2	tmp=tmp1+c2a2
state3	add out=a1g
state4	tmp1=add out-c1gtmp1
state5	tmp1=tmp+c3a3
state6	tmp=tmp1+c4a4
state7	a1=a2-c2tmp
state8	a2=a3-c3tmp1
state9	a3=a4-c4tmp
state10	add out=tmp1+d4tmp, read y
state11	add out=add out-y
state12	add out=add out+d1a1
state13	add out=add out+d2a2, a4=tmp
state14	Acc IN=add out+d3a3
state15	read x, goto state1

The elementary multiplication and addition operations, including the registers used for each step in Table 4.6 were firstly identified. Since this basic computation operation and register usage information had already been extracted, the register/output usage dependency was therefore known. According to this dependency information and the number of pipelines used in the multiplier, the execution sequence of the multiplier and adder had to be rearranged so that the pipeline of the multiplier was fully utilized. As a result, the throughput of the FE was thus maximized. The final FE internal execution table is tabulated in Table 4.7.

Genetic Operators. The genetic operation settings required to perform the selection, crossover, mutation, population size etc. are tabulated in Table 4.8.

System Performance. The speed of computation by FE is largely dependent upon the number of FEs being used. An independent study was carried out to investigate the relationship between the time taken to complete the number of iterations with the various numbers of FEs used. The result is summarized in Fig. 4.35. This undoubtedly demonstrates that the parallel architecture of the FE can greatly reduce computation time.

Table 4.7. The FE Internal Execution Sequence Table of the 4X4 IIR Lattice Filter

Mul A	Mul B	Mul P	Add In	Add Out	Add Op	Transfer
C1	TMP					
D1	TMP					
D2	TMP					
D3	TMP	C1G	'0'		Add	
D4	TMP	D1G	'0'	C1G	Add	
C1	A1	D2G	'0'	D1	Add	
C2	A2	D3G	'0'	D2	Add	
C3	A3	D4G	'0'	D3	Add	READ X
C4	A4	C1A1	X	D4	Add	
G	A1	C2A2	Add Out	TMP1	Add	
C1G	TMP1	C3A3	Add Out	TMP	Add	
C2	TMP	C4A4	Add Out	TMP1	Add	
C3	TMP1	A1G	'0'	TMP	Add	
C4	TMP	C1GTMP1	Add Out		Sub	
D4	TMP	C2TMP	A2	TMP1	Sub	READ Y
Y	'1'	C3TMP1	A3	A1	Sub	
D1	A1	C4TMP	A4	A2	Sub	
D2	A2	D4TMP	TMP1	A3	Add	A4=TMP
D3	A3	Y	Add Out		Sub	
C1	A1	D1A1	Add Out		Add	
C2	A2	D2GA2	Add Out		Add	
C3	A3	D3GA3	Add Out		Add	READ X
C4	A4	C1A1	X	Acc IN	Add	
.	.	C2A2	Add Out	TMP1	Add	
.	.	C3A3	Add Out	TMP	Add	
.	.	C4A4	Add Out	TMP1	Add	
.	.	.	.	TMP	.	

Table 4.8. Genetic Settings for ANC System

Population Size	30
Offspring generated per cycle	4
Selection	ranking
Crossover	Uniform crossover (rate=0.9)
Mutation	random mutation (rate = 0.1)

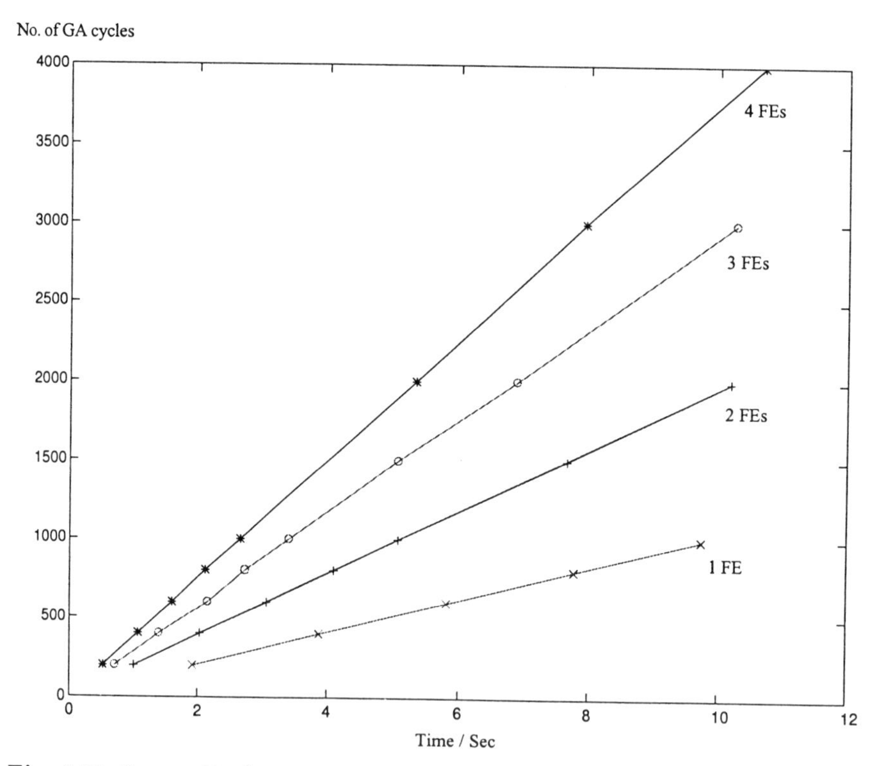

Fig. 4.35. System Performance

4.3 Case Study 3: GA in Automatic Speech Recognition

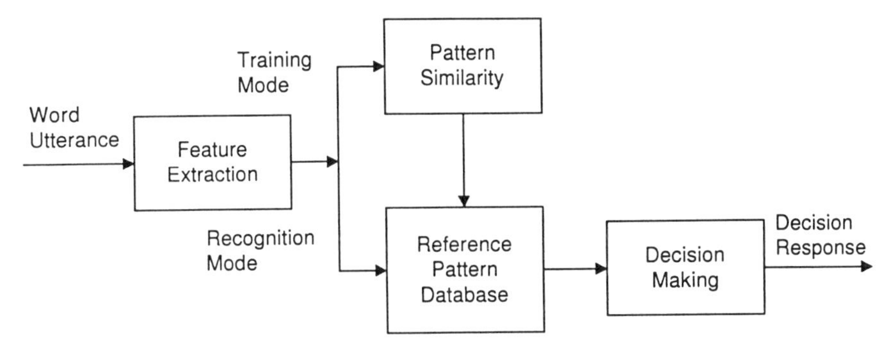

Fig. 4.36. Block Diagram of the Speech Recognition System

Another classic study to demonstrate the effectiveness of GA in engineering applications is the development of a Automatic Speech Recognition (ASR) system. Fig. 4.36 shows the overall ASR system that based on the template matching technique. It is composed of four main modules:

1. the feature extraction module;
2. the reference-pattern database;
3. the pattern similarity module and
4. the decision making module.

A typical ASR system usually runs in one of the two following modes:

1. the training mode; or
2. the recognition mode.

In either mode, the feature extraction module accepts the input speech utterance and divides it into blocks of equal numbers of samples called frames. Hence, a sequence of feature measurements is performed on each frame and the acoustic information is extracted within the frames. The extracted information forms a sequence of feature vectors which can be used to characterize the input 'speech patterns'.

Training mode – the output speech patterns from the feature extraction module are stored in a database as reference patterns for comparison purposes during the recognition phase. These speech patterns are then called 'reference patterns', and the database that is used to store the reference patterns is called the reference-pattern database.

Recognition mode – the output speech pattern of the feature extraction process is called the 'test pattern'. In this mode, the test pattern will not be stored in the reference pattern database, but will perform a similarity measurement between the test pattern and the reference patterns in the database. This is usually done by computing the spectral-distortion. A number of reference patterns that have distortion values less than a predefined value are selected and passed to the decision process module. According to some decision rules [115], the test pattern will be identified as the uttered word of the final selected reference pattern.

It is well known in speech recognition that individual speaking rates involve great variations [120]. These variations cause a nonlinear fluctuation, or timing difference of different utterances in the time-axis. This causes a major problem in computing the spectral-distortion measurement, and directly affects the performance of a template matching based ASR system. The simplest solution to this problem is to use a linear time normalization technique for the normalization of all speech patterns in a common time axis. However, experimental results show that these linear techniques are insufficient to deal with the highly nonlinear utterances.

Such a feature has been tackled by the use of a technique called dynamic time warping (DTW) [74, 125, 152, 156]. This is a dynamic programming-matching (DP-matching) [124] based nonlinear time-alignment scheme. DTW requires minimal system requirements, and is considered a faster training method but uses less searching times when compared with other available technique, such as the Hidden Markov Model. However, there are some drawbacks [74, 115] which may degrade the performance of DTW. The obstacles are largely classified as follows:

- the stringent requirement of the slope weighting function;
- the non-trivial finding of the K-best paths; and
- the relaxed endpoint constraint.

To overcome these problems, a stochastic method called Genetic Time Warping (GTW) is developed. GTW is a method that searches the warping paths, instead of using DP-matching. In this way, the problems that are confronted with DTW can be eliminated.

4.3.1 Warping Path

In order to formulate the GTW scheme, the warping path model should first be defined. Let $X = (x_1, x_2, \ldots, x_N)$ and $Y = (y_1, y_2, \ldots, y_M)$ represent two speech patterns, where x_i and y_i are the parametric vectors of the short-time acoustic features (any set of acoustic features can be used, as long as the distance measure for comparing pairs of feature vectors is known and its properties are well understood). We use i_x and i_y to denote the time indices of X and Y, respectively. The dissimilarity between X and Y is defined by a function of the short-time spectral distortions $d(x_{i_x}, y_{i_x})$, which will be denoted as $d(i_x, i_y)$ for simplicity, where $i_x = 1, 2, \ldots, N$ and $i_y = 1, 2, \ldots, M$.

A general time alignment and normalization scheme involves the use of two warping functions, ϕ_x and ϕ_y, which relate the indices of the two speech patterns, i_x and i_y, respectively. Both quantities are subjected to a common "normal" time axis k, i.e.,

$$i_x = \phi_x(k)$$
$$i_y = \phi_y(k) \quad \text{where} \quad k = 1, 2, \ldots, T$$

A global pattern dissimilarity measures $d_\phi(X, Y)$ based on the warping path $\phi = (\phi_x, \phi_y)$ (warping function pair) as the accumulated distortion over the entire utterance can thus be defined as follows:

$$D_\phi(X, Y) = \sum_{k=1}^{T} d\left(\phi_x(k), \phi_y(k)\right) m(k)/M_\phi \tag{4.47}$$

where $d(\phi_x(k), \phi_y(k))$ is a short-time spectral distortion defined for the $x_{\phi_x}(k)$ and $y_{\phi_y}(k)$; $m(k)$ is a non-negative (path) weighting coefficient; and M_ϕ is a (path) normalizing factor.

This clearly shows that an extremely large number of possible warping function pairs can be found. To extract some possible paths from this function is only possible if some kind of distortion measurement is used. One natural and popular choice [125] is to adopt a dissimilarity function $D(X, Y)$ as the minimum of $D_\phi(X, Y)$ over a whole range of possible paths and this is achieved by minimizing Eqn. 4.47 such that

$$D(X, Y) = \min_\phi D_\phi(X, Y) \tag{4.48}$$

where the warping function (path) ϕ must satisfy a set of conditions [115, 156].

By the same token, the warping path ϕ must also obey the properties of the actual time-axis. In other words, when the mapping of the time axis of a pattern X onto the time axis of a pattern Y is necessary, while the linguistic structures of the speech patterns X must also be preserved. These linguistic structures are characterized by the continuity, temporal order, local continuity and other acoustic information of the speech signals.

For a realistic warping path ϕ, the following criteria must be satisfied:

- the endpoint constraint;
- the local monotonic constraint;
- the local continuity;

– the slope constraint; and
– the allowable region.

Endpoint Constraint. The uttered signal in the signal stream must have well-defined endpoints that mark the beginning and the ending frames of the signal for recognition, i.e.,

$$\begin{aligned} \text{beginning point:} \quad & \phi_x(1) = 1, \quad \phi_y(1) = 1 \\ \text{ending point:} \quad & \phi_x(T) = N, \quad \phi_y = (T) = M \end{aligned} \tag{4.49}$$

It should be noted that the endpoints are easily disrupted by background noise. As a result, it can cause inaccurate endpoint estimation. Hence, the endpoint constraints are usually relaxed, i.e.,

$$\begin{aligned} \text{beginning point:} \quad & \phi_x(1) = 1 + \Delta x, \quad \phi_y(1) = 1 + \Delta y \\ \text{ending point:} \quad & \phi_x(T) = N - \Delta x, \quad \phi_y(T) = M - \Delta y \end{aligned} \tag{4.50}$$

Local Monotonicity. The temporal order is a special feature of acoustic information in speech patterns. Therefore, the warping path cannot be traversed reversibly. A local monotonic constraint should be imposed to prevent such occurrence. This can be done by restricting ϕ, i.e.

$$\begin{aligned} & \phi_x(k+1) \geq \phi_x(k) \\ & \phi_y(k+1) \geq \phi_y(k) \end{aligned} \tag{4.51}$$

Local Continuity. This property ensures the correct time alignment and maintains the integrity of the acoustic information. The local continuity constraint imposed on ϕ is as follows:

$$\begin{aligned} & \phi_x(k+1) - \phi_x(k) \leq 1 \\ & \phi_y(k+1) - \phi_y(k) \leq 1 \end{aligned} \tag{4.52}$$

Slope Constraint. Neither too steep nor too gentle a gradient should be allowed in ϕ, otherwise unrealistic time-axis warping may appear. A too steep or too gentle gradient in ϕ implies comparison of a very short pattern with a relatively long pattern. Therefore, the weighting function $m(k)$ in Eqn. 4.47 controls the contribution of each short-time distortion $d(\phi_x(k), \phi_y(k))$ which can be used to eliminate the undesired time-axis warping.

Slope constraint is recognized as a restriction on the relation of several consecutive points on the warping function rather than a single point. As indicated in Figure 4.37a, if ϕ moves along the horizontal m times, then ϕ must move along in a diagonal direction for at least n times before stepping against the horizontal direction. The effective intensity of the slope constraint can be measured by the parameter $P = n/m$. H. Sakoe and S. Chiba [125] suggested four types of slope constraint with different values of P. This can be seen in Fig. 4.37c.

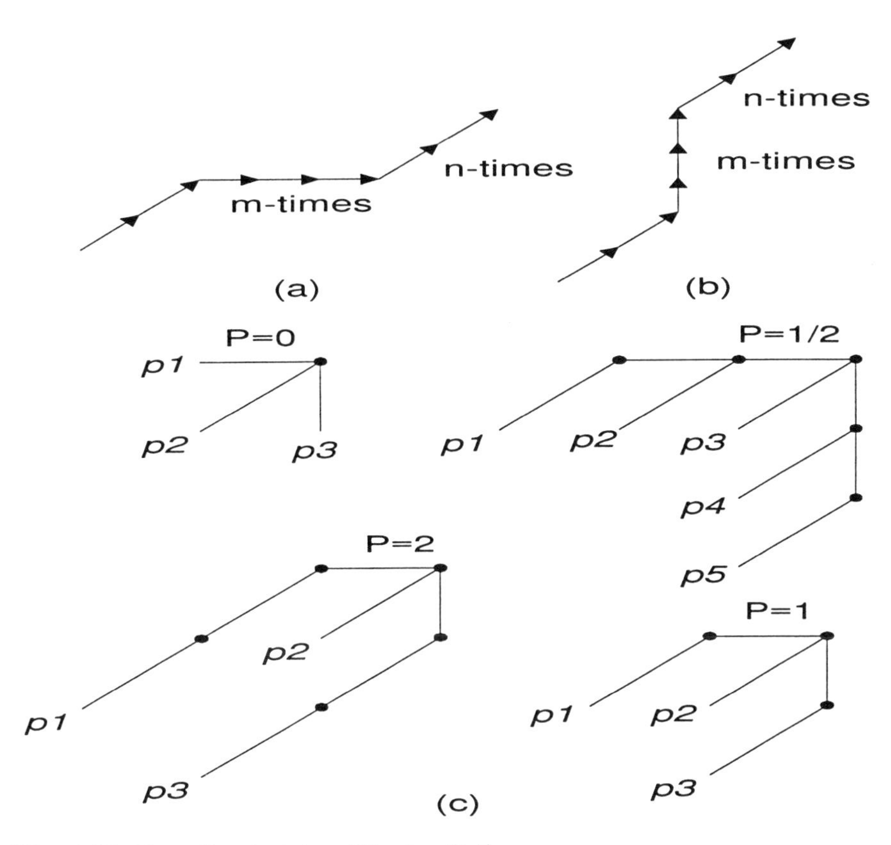

Fig. 4.37. Slope Constraint on Warping Path

Allowable Region. Because of slope constraints, certain portions of (i_x, i_y) are excluded from the searching region. The allowable region can be defined by two parameters: Q_{max} and Q_{min} the maximum slope and minimum slope of searching steps in the DP-searching path, respectively. For example, according to Fig. 4.37c, the slope constraint with $P = 1/2$ has $Q_{max} = 3$ and $Q_{min} = 1/3$, where $p1$ to $p5$ are the allowable paths that satisfy the slope constraint, where $p1$ has the minimum slope of 1/3 while $p5$ has the maximum slope of 3.

The allowable regions of Q_{max} and Q_{min}, can be defined as follows:

$$1 + \frac{\phi_x(k) - 1}{Q_{max}} \le \phi_y(k) \le 1 + Q_{max}\left(\phi_x(k) - 1\right) \tag{4.53}$$

$$M + Q_{max}\left(\phi_x - N\right) \le \phi_y(k) \le M + \frac{\phi_x(k) - N}{Q_{max}} \tag{4.54}$$

Eqn. 4.53 specifies the range of the points in the (i_x, i_y) plane that can be reached from the beginning point $(1, 1)$. Similarly, Eqn. 4.54 specifies the range of points which have legal paths to the ending point (N, M). The overlapped region of Eqns. 4.53 and 4.54 form the allowable region. It should be noted that the allowable region may not exist if the regions described by Eqns. 4.53 and 4.54 are totally disjointed, i.e. $(M-1)/(N-1) > Q_{max}$ or $(M-1)/(N-1) < Q_{min}$, because it is unrealistic to compare two utterances that have very large time variations.

To solve Eqn. 4.48, the searching method DTW was used. DTW uses the dynamic programming technique [124] to search the optimal warping path ϕ for the dissimilarity function $D(X, Y)$. This technique had been evaluated by many research groups [74, 152, 156] such that a recognition accuracy factor as high as 99.8% can be maintained [125]. However, due to the recursive approach of the dynamic programming technique, some restrictions on DTW, such as the stringent rule on slope weighting and the non-trivial finding of the K-best paths do exist. These restrictions have been raised by L.R. Rabiner and B.H. Juang [115].

4.3.2 Implementation of Genetic Time Warping

Having established the essential arguments for the formulations of the time warping problem, the actual implementation procedure may proceed. Considering the intrinsic properties of GA, a number of approaches may be adopted. The following subsections provide a detailed account of the methods used and comparison have been made to iron out their differences.

Genetic Time Warping. This is an obvious approach for implementing GA to solve the time warping problem. It uses a conventional GA for implementation. Here, the optimal warping path searching problem has to be mapped into the GA domain. These include the following considerations:

- a mechanism to encode the warping path ϕ as a chromosome;
- a fitness function to evaluate the performance of ϕ;
- a selection mechanism;
- genetic operators (crossover and mutation).

Encoding Mechanism. In order to proceed with GTW, the warping path must be encoded in the form of a chromosome. Consider that a warping path $\phi' = (\phi'_x, \phi'_y)$ is represented by a sequence of points in the (i_x, i_y) plane, i.e.,

$$\phi' = \left(\phi'_x(1), \phi'_y(1)\right) \left(\phi'_x(2), \phi'_y(2)\right) \ldots \left(\phi'_x(T), \phi'_y(T)\right) \tag{4.55}$$

Eqn. 4.55 must satisfy the allowable region and local constraints described in Sect. 4.3.1. The order of points must move along the allowable paths which are restricted by the slope constraint. Thus, it is preferable to express the warping function ϕ by a sequence of sub-paths. For example, the sub-paths $p1, p2, \ldots, p5$ in Fig. 4.37(c) are encoded as $1, 2, \ldots, 5$. Therefore, the warping function ϕ can be expressed as follows:

$$\phi = (p_1)(p_2)\ldots(p_L) \tag{4.56}$$

with initial points (i_0, i_0) and p_n being encoded as the allowable sub-paths.

There are several parameters to be initialized, such as the allowable region and local constraints of the warping function. For example, Δx and Δy in Eqn. 4.50 define the degree of relaxation of the endpoints, P defines the degree of slope constraint and Q_{max} and Q_{min} in Eqns. 4.53 and 4.54 define the dimensions of the allowable region. Once the encoding scheme and the initialization of the system parameters are completed, the population pool can be generated. The initialization procedures are summarized as follows:

1. Randomly select a beginning point;
2. Randomly select a sub-path;
3. Calculate the position of the ending point of the selected sub-path;
4. If the ending point of the selected sub-path falls outside the allowable region, then goto step 2;
5. Encode the gene as the selected sub-path and the absolute position of the ending point of the selected sub-path;
6. If the global ending points (n, m), where $N - \Delta x \leq n; \leq N$ and $M - \Delta y \leq m \leq M$, are not reached, then goto step (2); and
7. Repeat all steps until entire population is initialized.

Fitness Function. Eqn. 4.47 is a distortion measurement between the two utterances X and Y. This provides the mechanism to evaluate the effectiveness of the warping function. However, the range of the distortion values calculated by Eqn. 4.47 can be unbounded. A fitness function is designed to normalize Eqn. 4.47 to a range from 0 to 1. The normalized value of Eqn. 4.47 is the fitness of the warping path function, which is used by the selection mechanism to evaluate the value of the "survival-of-the-fittest" of the warping function in subsequent generations. The procedures to calculate the fitness values are as follows:

1. Calculate distortion values d_n of each warping function in the population by Eqn. 4.47, i.e. $d_n = d_{\phi_n}(X, Y)$;
2. Find the maximum distortion value d_{max} in the population, i.e. $d_{max} = \max(d_1, d_2, \ldots, d_s)$;
3. Calculate the absolute differences dif_n between d_{max} and each d_n, i.e. $dif_n = d_{max} - d_n$;
4. Calculate the summation T of all differences calculated in step (3), i.e. $T = \sum dif_n$; and
5. Calculate fitness values f_n of each warping function by equation $f_n = dif_n / T$.

Selection. The selection procedure is modelled after nature's "survival-of-the-fittest" mechanism. Fitter solutions survive and weaker ones die. After selection, the fitter solutions produce more offspring and thus have a higher chance of surviving in subsequent generations.

GTW uses the Roulette wheel selection scheme as its selection mechanism. In this type of selection, each solution is allocated a sector of the Roulette wheel with the angle subtended by sector at the centre of the wheel, which is equal to 2π multiplies by the fitness value of the solution. A solution is selected as an offspring if a randomly-generated number in the range 0 to 2π falls into the sector corresponding to the solution. The algorithm selects solutions in this manner until the entire population of next generation has been produced. The procedures used to implement the Roulette wheel selection scheme are outlined as follows:

1. Create an array *Sector* with $S-1$ real numbers where S is the population size, i.e. real $Sector[1..S-1]$;
2. Set *1*-st item of $Sector = 3600 \times$ fitness value of 1st warping path of the population, i.e. $Sector[1] = 3600 \times f_1$;
3. Iterate n from 2 to $S-1$, set n-th item of $Sector =$ (n-1)-th item of $Sector + 3600 \times$ fitness value of n-th warping path in the population, i.e. in each iteration $Section[n] = Section[n-1] + 1000 \times f_n$;
4. Randomly select a number p from 0 to 3600;
5. Find index i of *Sector* such that $Sector[i]$ is minimum in *Sector* and $Sector[i] \geq p$. If i does not exist, set $i = S$. Then select i-th warping

path of the current population to the population of the next generation; and
6. Repeat from step (4) until entire population of next generation has been selected.

Crossover. The crossover procedure implements the exchange mechanism between two parent chromosomes. Crossover between selected fitter chromosomes in the population possibly reproduces a more optimized solution. The crossover rate is controlled by a probability factor which controls the balance between the rate of exploration, the new recombined building block and the rate of disruption of good individuals.

Since each chromosome of GTW represents a continuous warping path, arbitrary exchange between two chromosomes may generate two discontinuous warping paths. GTW has to ensure that the newly generated offspring remain in a continuous form. The procedures for crossover between two chromosomes are outlined as follows:

1. Randomly choose two warping paths A and B from the population;
2. Randomly generate a number p_{cross} from 0 to 1. If $p_{cross} >$ crossover rate then use A as the offspring and finish;
3. Randomly choose a gene g_s from A, use the ending point stored in g_s as cross point s;
4. Search a gene g_e from B which has a ending point e such that point s can move to point e along an allowable path p_c. If no such gene exists then use A as the offspring and finish;
5. The offspring will be composed of two parts: *1*-st part is the segment of A from the *1*-st gene to g_s, *2*-nd part is the segment of B from the g_e to the last gene. Modify the allowable path in g_e of the offspring to p_c.

Mutation. The process of mutation randomly alters the genes of the chromosomes and acts in the role of restoring lost genetic material that has not been generated in the population during the initialization procedure. Because new genetic information will not be generated by the crossover procedure, the mutation procedure becomes an important mechanism to explore new genetic information.

Mutation is used to provide a random search capability for GA. This action is necessary for solving practical problems that arise with multimodal functions. The mutation rate is also controlled by a probability factor which controls the balance between random searching and the rate of disruption of good individuals. It should be noted that the random alteration of genes during mutation may result in a discontinuous warping path. So a special treatment of mutation in GTW is used to avoid this situation. The mutation procedures are summarized as follows:

1. Randomly choose a warping path A from the population;
2. Randomly generate a number p_{mut} from 0 to 1. If $p_{mut} >$ mutation rate then use A as the offspring and stop;
3. Randomly choose two genes g_s and g_e from A where g_s is positioned at a position ahead of g_e. The ending points stored in g_s and g_e marked as s and e, respectively.
4. Initialize a warping path p_m between points s and e;
5. The offspring will be generated by replacing the genes of A between g_s and g_e with p_m.

When all the GA operations are defined for the optimal time warping path searching problem, the evolutionary cycle of GA in Fig. 4.38 can then be started. It can be seen that the GA-based time normalization technique can solve the above mentioned problems in Sect. 4.3.1 for the DTW.

A number of points need to be clarified in order to bring about the best use of GTW:

1. the population size or the number of evaluations is constant in the GTW. This implies that the computational requirement is independent of the degree of endpoint relaxations;
2. the DTW solved by the dynamic programming technique is unable to compute the M_ϕ dynamically due to its recursive operation. This usually restricts the M_ϕ in Eqn. 4.47 to a constant;
3. GTW considers the solution of the warping function on a whole-path basis rather than a sub-path by sub-path basis. The M_ϕ used to evaluate the warping function can be obtained directly in each fitness evaluation;
4. the constraints of the slope weighting function $m(k)$ are relaxed and can be arbitrarily chosen in this case, such that M_ϕ need not to be a constant for all possible paths; and
5. the speech pattern distortions calculated in DTW are the minimum values of a number of traversed paths, and the Backtrack technique [115] can only be used to compute a single optimal path so that it is difficult to obtain the second-best and the third-best warping paths.

The GA operates on a pool of population and all the warping paths are stored as chromosomes in the pool. Therefore, warping paths can all be evaluated independently so that the K-best paths can be obtained naturally and without extra computational requirements.

Genetic Time Warping with Relaxed Slope Weighting Function. Sect. 4.3.1 has pointed out that the comparison between two utterance with large timing different is unrealistic. This problem can be alleviated by introducing a normalized slope weighting function on $m(k)$. This is possible when M_ϕ in Eqn. 4.47 is defined as:

$$M_\phi = \sum_{k=1}^{T} m(k) \tag{4.57}$$

The computation of M_ϕ can be clumsy for DTW particularly when $m(k)$ is varied dynamically. Whereas for the case in GTW, each path is coded as a chromosome, then the computation of M_ϕ presents no problem in the GTW formulation. With such an unique property, the definition of $m(k)$ is therefore relaxed and can be chosen arbitrarily. In this way, a GTW scheme with a relaxed slope weight function (GTW-RSW) can thus be performed.

Hybrid Genetic Algorithm. The GTW described above will produce results, and it is a well known fact that the classic GAs do have the inclination to search optimally over a wide range of the dynamic domain. However, they also suffer from being slow in convergence. To enhance this searching capability and improve the rate of convergence, problem-specific information is desirable for GA so that a hybrid-GA structure is formed. In the present hybrid-GA formulation, we add problem-specific knowledge to the crossover operator so that reproduction of offspring that possesses higher fitness values is realized.

In this hybrid-GTW scheme, the hybrid-crossover genetic operator is proposed. The hybrid-crossover operator is similar to the original crossover operator whose procedure is listed as follows:

1. Randomly select two chromosomes A and B and perform the normal crossover operation. An offspring C is produced;
2. Swap chromosomes A and B and perform the crossover procedures again. Another offspring D is produced; and
3. Instead of putting the offspring back into the population, a discrimination process is executed, such that the best chromosomes among A, B, C, and D will be put back into the population pool.

The experimental results of the hybrid approach of GTW are shown in Sect. 4.3.3, which indicates that the hybridized GTW achieves better results than the tradition GTW using the same number of generations.

Parallel Genetic Algorithm. Sect. 3.1 has already described the properties of intrinsic parallelism in GA. The same technique may apply here. In this application, there will be more chromosomes (warping paths) to be examined. However, the implementation of GA parallelism for GTW is not straight forward due to the need (in the selection steps) for a global statistics calculation which results in higher communication costs.

In order to reduce these costs, global selection is therefore avoided, but only the K-best solutions of each processor are considered for communication. The operational procedures are described as follows:

1. Each worker processor extracts K best solutions and $K \times N$ bad solutions from its population where N is the number of processors in the system provided that $K \times N <$ size of the population;
2. Each worker processor passes the K-best solutions to the farmer processor;
3. The farmer processor puts all received solutions into a group of $K \times N$ solutions and sends the group back to every processor;
4. Each worker processor replaces $K \times N$ bad solutions in its population by the received solutions.

To realize this particular application, a dedicated parallel processing hardware prototype is built for evaluating the parallel-GTW performance. This architecture is shown in Fig. 4.38. A global 16-bit communication bus, acts as the communication channel between the four processors being used. All communication and data transfer between the four processors are supported by this channel. A host computer with an eight-bit communication bus forms the communication channel between the host computer and the four processors. All reference and test speech signal patterns are downloaded from the host computer to each processor via this channel.

This is homogenous parallel system in which the four processor boards (PE #1 to PE #4) are identical. Each board has a DSP56001 digital signal processor, a 4Kword (16 bit word) dual-port memory used as shared memory, a 32Kword local memory for each PE, a bridge (buffer) between the global communication bus and the memory bus of the DSP56001, and the host port of the DSP56001 are all connected to the host computer's communication bus. The performance of this scheme is shown in the next section.

4.3.3 Performance Evaluation

To evaluate the performance of the above-mentioned schemes for the time warping problem, experimental results of the following five algorithms have been obtained:

1. Dynamic Time Warping (DTW)
   The warping paths obtained by DTW used the dynamic-programming searching method proposed by L.R. Rabiner [115].
2. Genetic Time Warping (GTW)
   GTW used the traditional Genetic Time Warping technique described in Sect. 4.3.2. GTW used the $m(k)$ as defined in Eqn. 4.59, i.e. the normalization factor M_ϕ in Eqn. 4.47 must be constant for all warping paths.
3. Genetic Time Warping with Relaxed Slope Weighting function (GTW-RSW)
   GTW-RSW is the same as GTW except that the slope weighting function

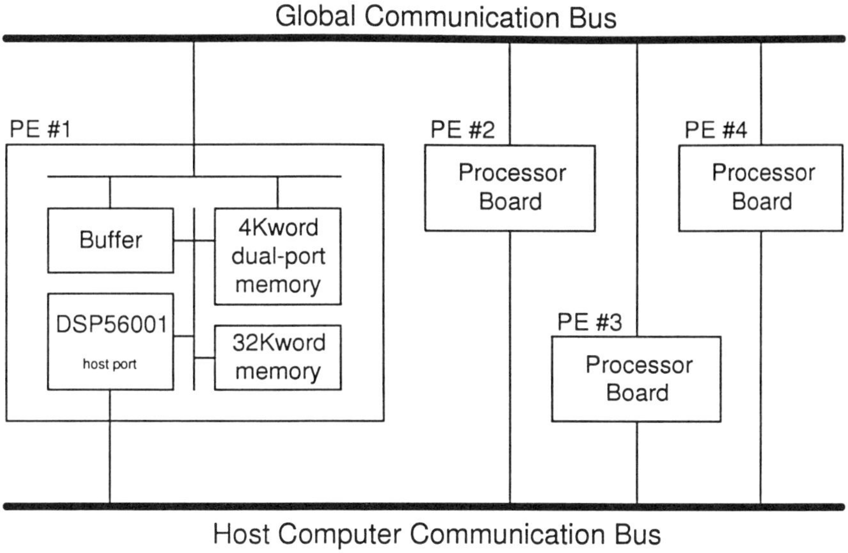

Fig. 4.38. Hardware Architecture of the Parallel-GTW

$m(k)$ used by GTW-RSW is relaxed. This means that the values of $m(k)$ can be arbitrarily chosen or $m(k)$ relaxed. M_ϕ can be varied for different warping paths.

4. Hybrid Genetic Time Warping (hybrid-GTW)
   Hybrid-GTW is the same as GTW-RSW except that it uses the hybrid-crossover operator described in Sect. 4.3.2 instead of traditional crossover operator.
5. Parallel Genetic Time Warping (parallel-GTW)
   Unlike the above four algorithms which are executed on a single processor platform, the parallel-GTW is executed on a multiprocessor platform described in Sect. 4.3.2. Each processing element (PE) in the platform executes a GTW-RSW algorithm. After a generation cycle, the best 10 chromosomes in each PE are broadcast to three other PEs for the next evolution.

A database of 10 Chinese words spoken by two different speakers was used with 100 utterances for each word. Each utterance was sampled at 8.0KHz rate, 8-bit digitized and divided into frames of 160 samples. Ten-order cepstral analysis was applied as the feature measurement for the feature extractions. The initial and final endpoints for each word were determined by a zero cross rate and energy threshold. The short-time spectral distortion measurement is

$$d(a_R, a_T) = \sum_{i=1}^{10} |a_{R_i} - a_{T_i}| \tag{4.58}$$

where a_R and a_T are the short-time spectral feature vectors of reference and test patterns, respectively.

For each of the 10 words, 80 arbitrarily chosen utterances act as the reference patterns while the remaining 20 utterances are the samples for test patterns.

Each warping path in our experiments has five relaxed beginning points and ending points, i.e. the Δx and Δy in Eqn. 4.50. The slope constraint for P was set to 1/2 (as shown in Fig. 4.37(c)). The allowable region was defined as $Q_{max} = 3$ and $Q_{min} = 1/3$ in Eqns. 4.53 and 4.54. The following slope weighting function $m(k)$ for DTW has the following form:

$$m(k) = \phi_x(k) - \phi_x(k-1) + \phi_y(k) - \phi_y(k-1) \tag{4.59}$$

Table 4.9, summarizes the $m(k)$s used for the allowable step p_n for the DTW, while Table 4.10 shows the p_n for GTW, GTW-RSW, hybrid-GTW and parallel-GTW. The GA operational parameters are tabulated in Table 4.11

Table 4.9. Slope Weighting Function Used in DTW

Allowable path used	$m(k)$ in Eqn. 4.47
p1	4
p2	3
p3	2
p4	3
p5	4

Table 4.10. Slope Weighting Function Used in GTW, GTW-RSW, Hybrid-GTW and Parallel-GTW

Allowable path used	$m(k)$ in Eqn. 4.47
p1	5
p2	3
p3	1
p4	3
p5	5

Tables 4.12-4.16 list the results of the five experiments. The results are given in terms of M_s, δ_s, M_d and δ_s in which the mean distortions (Eqn. 4.47) of the same words, the standard deviations of distortions of the same words, the mean distortions of different words and the standard deviations

Table 4.11. Genetic Parameter for GTW

Population Size	40
Crossover Rate	0.6
Mutation Rate	0.03333
Maximum Generation	40

of different words, are given respectively.

Table 4.12. Experimental Results of DTW

word	M_s	δ_s	M_d	δ_d
1	0.757	1.050	4.614	28.734
2	0.715	0.998	5.287	40.170
3	0.832	1.167	5.195	37.687
4	0.610	0.874	7.239	63.138
5	0.800	1.123	4.562	24.323
6	0.802	1.115	4.352	20.917
7	0.785	1.105	6.106	45.917
8	0.915	1.289	4.364	24.275
9	0.726	1.012	3.924	16.714
10	0.792	1.102	4.104	19.41

Table 4.13. Experimental Results of GTW

word	M_s	δ_s	M_d	δ_d
1	1.125	1.670	5.475	38.98
2	0.959	1.362	6.136	52.335
3	1.322	2.142	5.985	47.339
4	0.789	1.101	8.202	79.839
5	1.202	1.861	5.443	33.958
6	1.244	1.944	5.210	29.477
7	1.092	1.638	7.024	59.473
8	1.328	2.143	5.107	31.867
9	1.088	1.603	4.629	22.954
10	1.321	2.133	4.954	27.30

To assess the performance of the ASR system, a single value of the mean distortion M_s is not enough for classification. This also compounds the fact that a variation of the slope weighting functions $m(k)$ (in Eqn. 4.47) has been used in the experiments.

Table 4.14. Experimental Results of GTW-RSW

word	M_s	δ_s	M_d	δ_d
1	1.017	1.654	5.537	36.087
2	0.939	1.326	6.263	49.457
3	1.257	2.134	6.310	37.238
4	0.773	1.078	8.319	71.919
5	1.002	1.859	5.305	32.118
6	1.182	1.950	5.496	28.320
7	1.085	1.618	7.517	50.716
8	1.307	2.090	5.670	28.980
9	0.982	1.622	5.389	21.080
10	1.106	2.156	4.963	24.982

Table 4.15. Experimental Results of Hybrid-GTW

word	M_s	δ_s	M_d	δ_d
1	0.911	1.638	5.599	33.191
2	0.909	1.290	6.390	46.579
3	1.192	2.126	6.635	27.137
4	0.757	1.055	8.436	63.999
5	0.802	1.857	5.167	30.278
6	1.120	1.956	5.782	27.163
7	1.078	1.598	8.010	41.959
8	1.286	2.037	6.233	26.093
9	0.876	1.641	6.149	19.206
10	0.891	2.179	4.972	22.657

Table 4.16. Experimental Results of Parallel-GTW

word	M_s	δ_s	M_d	δ_d
1	0.926	1.647	5.612	33.197
2	0.922	1.305	6.399	46.588
3	1.205	2.144	6.637	27.150
4	0.759	1.071	8.447	64.006
5	0.814	1.866	5.174	30.292
6	1.130	1.965	5.800	27.181
7	1.086	1.611	8.014	41.968
8	1.295	2.038	6.240	26.108
9	0.892	1.652	6.164	19.211
10	0.903	2.198	4.989	22.662

Furthermore, the recognition rates of both the DTW and GA approaches were very close and found to be almost identical. Therefore, a more reasonably accurate measurement for the assessment is to use the absolute difference between the M_s and M_d, i.e. $|M_s - M_d|$. This provides discriminating abilities for recognizing confused utterances, particularly for utterances that have similar acoustic properties.

In this case, a lower value in $|M_s - M_d|$ implies that the ASR system has a poor ability to identify confused utterances, while a higher $|M_s - M_d|$ value implies that the ASR system has a high level of confidence in recognizing confused utterances. The results of $|M_s - M_d|$ for the five experiments are tabulated in Fig. 4.39.

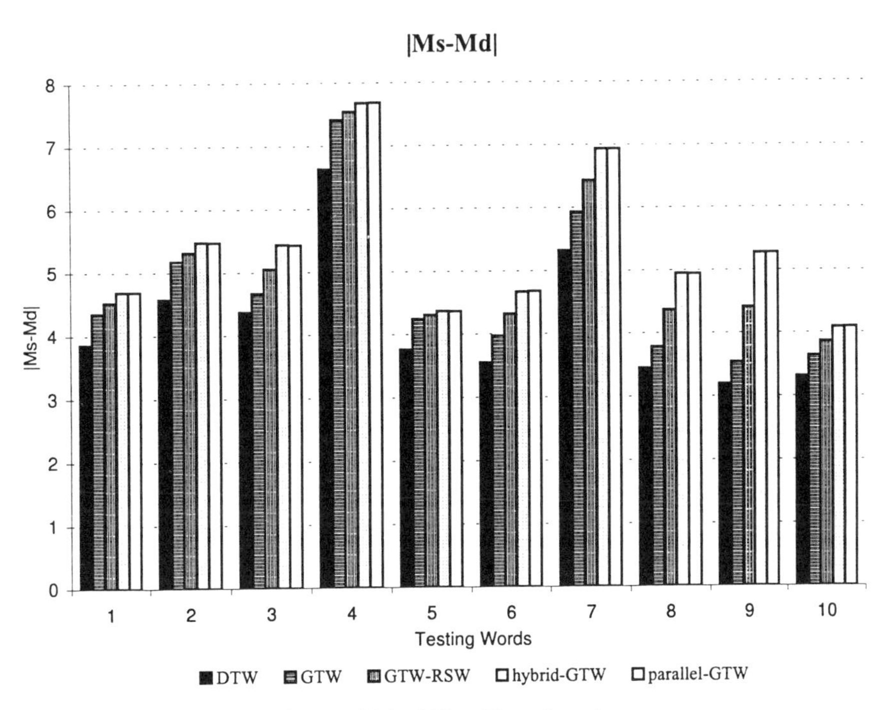

Fig. 4.39. The Chart of $|M_s - M_d|$ of Five Experiments

It can be clearly shown in Fig. 4.39 that all the algorithms using the GA technique have higher values of $|M_s - M_d|$ than those of DTW, and therefore a higher discrimination ability than the conventional DTW.

It is also expected that the hybrid and parallel approaches of the GTW should have a faster convergence time. Fig. 4.40 summarizes the results of M_s for GTW-RSW, hybrid-GTW and parallel-GTW. As tabulated in the table, the hybrid-GTW and the parallel-GTW have smaller values of M_s than the GTW-RSW with the same number of generations in the evolutional cycle. This implies that the hybrid-GTW and the parallel-GTW have a faster convergence time than the GTW-RSW and further verifies that the use of hybrid-GTW and parallel-GTW can speed up the searching process.

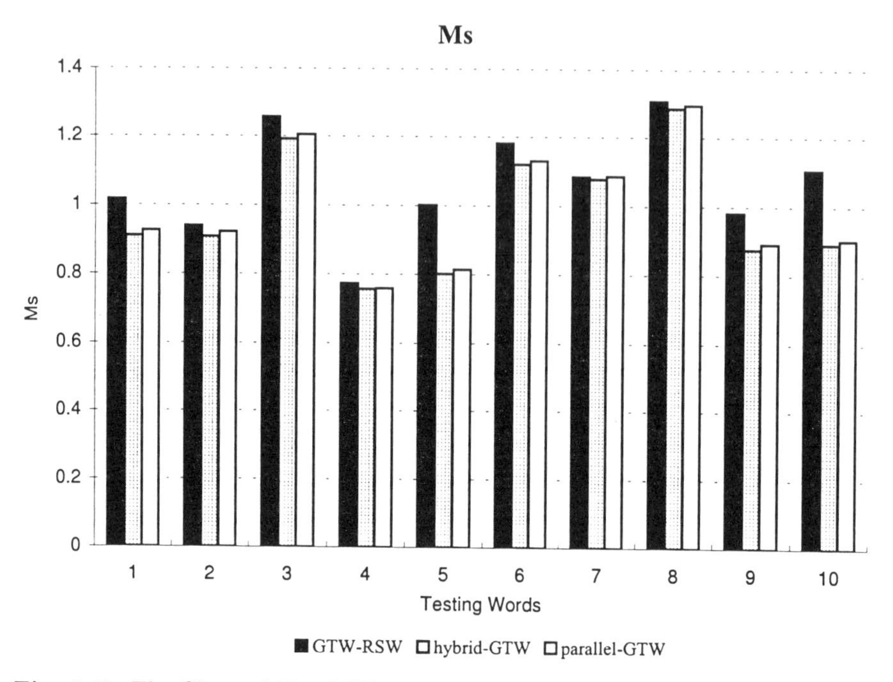

Fig. 4.40. The Chart of Ms of GTW-RSW, Hybrid-GTW and Parallel-GTW

CHAPTER 5
HIERARCHICAL GENETIC ALGORITHM

Thus far, the essence of the GA in both theoretical and practical domains has been well demonstrated. The concept of applying a GA to solve engineering problems is feasible and sound. However, despite the distinct advantages of a GA for solving complicated, constrained and multiobjective functions where other techniques may have failed, the full power of the GA in engineering application is yet to be exploited and explored.

To bring out the best use of the GA, we should explore further the study of genetic characteristics so that we can fully understand that the GA is not merely a unique technique for solving engineering problems, but that it also fulfils its potential for tackling scientific deadlocks that, in the past, were considered impossible to solve. In this endeavour, we have chosen as our target an examination of the biological chromosome structure. It is acknowledged in biological and medical communities that the genetic structure of a chromosome is formed by a number of gene variations, that are arranged in a hierarchical manner. Some genes dominate other genes and there are active and inactive genes. Such a phenomenon is a direct analogy to the topology of many engineering systems that have yet to be determined by an adequate methodology.

In light of this issue, a reprise of biological genetics was carried out. A method has been proposed to emulate the formulation of a biological DNA structure so that a precise hierarchical genetic structure can be formed for engineering purposes. This translation from a biological insight enables the development of an engineering format that falls in-line with the usual GA operational modes of action. The beauty of this concept is that the basic genetic computations are maintained. Hence, this specific genetic arrangement proves to be an immensely rich methodology for system modelling, and its potential uses in solving topological scientific and engineering problems should become a force to be reckoned with in future systems design.

This chapter outlines this philosophical insight of genetic hierarchy into an engineering practice in which all the necessary techniques and functionalities are described. The achievements that can be derived from this method for solving typical engineering system topology designs are given in the following chapters.

5.1 Biological Inspiration

5.1.1 Regulatory Sequences and Structural Genes

The biological background of the DNA structure has already been given in Chap. 1. An end product, generally known as a polypeptide, is produced by the DNA only when the DNA structure is experiencing biological and chemical processes. The genes of a complete chromosome may be combined in a specific manner, in which there are some active and inactive genes. It has become a recognized fact that there are 4,000 genes in a typical bacterial genome, or an estimated 100,000 genes in a human genome. However, only a fraction of these genes in either case will be converted into an amino acid (polypeptide) at any given time and the criteria for a given gene product may change with time. It is therefore crucial to be able to regulate the gene expression in order to develop the required cellular metabolism, as well as to orchestrate and maintain the structural and functional differences between the existing cells.

From such a genetic process, the genes can thus be classified into two different types:

- regulatory sequences (RSs) and
- structural genes (SGs)

The SGs are coded for polypeptides or RNAs, while the RSs serve as the leaders that denote the beginning and ending of SGs, or participate in turning on or off the transcription of SGs, or function as initiation points for replication or recombination. One of the RSs found in DNA is called the promoter, and this activates or inactivates SGs due to the initialization of transcription. This initialization is governed by the Trans-acting Factor (TAF) [33, 78, 96] acting upon this sequence in the DNA. The transcription can only be taken place if a particular TAF is bound on the promoter. A polypeptide is then produced via a translation process in which the genetic message is coded in mRNA with a specific sequence of amino acids. Therefore, a hierarchical structure is obtained within a DNA formation that is depicted in Fig. 5.1.

5.1.2 Active and Inactive Genes

One of the most surprising discoveries in the founding of molecular biology was that active and inactive genes exist in the SGs. The active genes are

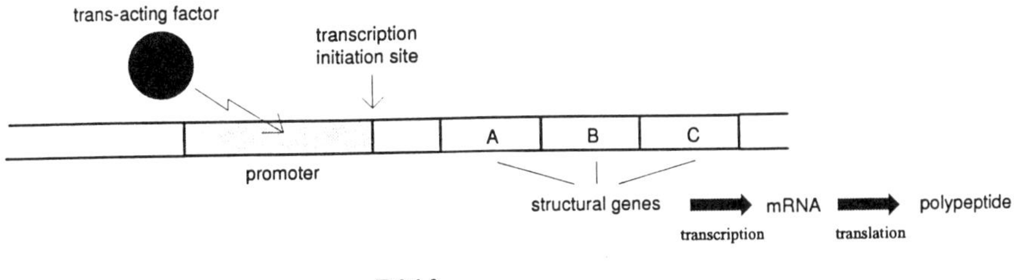

Fig. 5.1. Trans-acting Factor Bound on Promoter for the Initiation of Transcription

separated into non-contiguous pieces along the parental DNA. The pieces that code mRNA are referred to as exons (active genes) and the non-coding pieces are referred as introns (inactive genes). During transcription, there is a process of splicing, Fig. 5.2 so that the final messenger RNA, which contains the exons only, are formed.

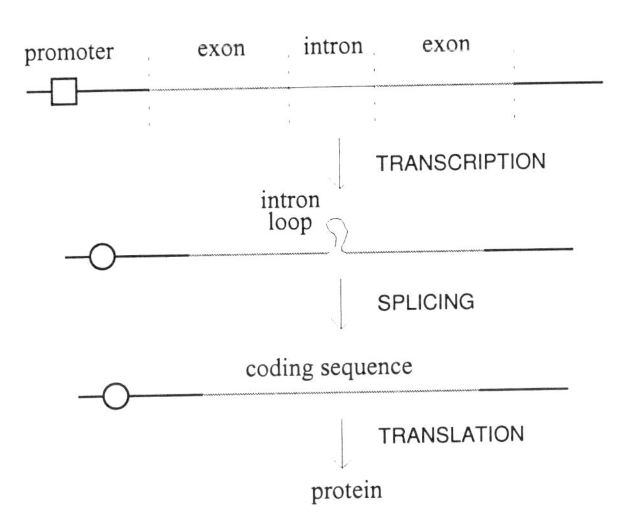

Fig. 5.2. Splicing

5.2 Hierarchical Chromosome Formulation

It is the existence of the promoter together with the commanding signals of TAF that ignites the innovated introduction of hierarchical formulation of the chromosome for GA. This chromosome can be regarded as the DNA that has already been described, but consists of the parametric genes (analogy to structural genes in DNA) and the control genes (analogy to regulatory sequences in DNA). This architecture is very similar to that shown in Fig. 5.1.

To generalize this architecture, a multiple level of control genes are introduced in a hierarchical fashion as illustrated in Fig. 5.3. In this case, the activation of the parametric gene is governed by the value of the first-level control gene, which is governed by the second-level control gene, and so on.

To indicate the activation of the control gene, an integer "1" is assigned for each control gene that is being ignited where "0" is for turning off. When "1" is signalled, the associated parameter genes due to that particular active control gene are activated in the lower level structure. It should be noticed that the inactive genes always exist within the chromosome even when "0" appears. This hierarchical architecture implies that the chromosome contains more information than that of the conventional GA structure. Hence, it is called Hierarchical Genetic Algorithm (HGA).

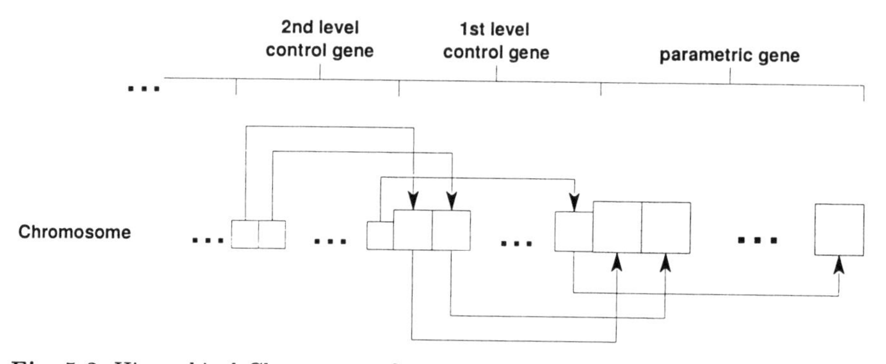

Fig. 5.3. Hierarchical Chromosome Structure

The use of the HGA is particularly important for the structure or topology as well as the parametric optimization. Unlike the set-up of the conventional GA optimization, where the chromosome and the phenotype structure are assumed to be fixed or pre-defined, HGA operates without these constraints. To illustrate this concept further, the following example is used to demonstrate the functionality of the HGA for engineering applications.

Example 5.2.1. Consider a chromosome formed with 6-bit control genes and 6-integer parametric genes as indicated by Fig. 5.4:

The length of X_A and X_B are 4 and 2, respectively, which means that the phenotype in different lengths is available within the same chromosome formulation. Then, the HGA will search over all possible lengths including the parameters so that the final objective requirement is met. Moreover, the hierarchical levels can be increased within the chromosome to formulate a multilayer chromosome. This is shown in Fig. 5.5 where a three-level gene structure is represented.

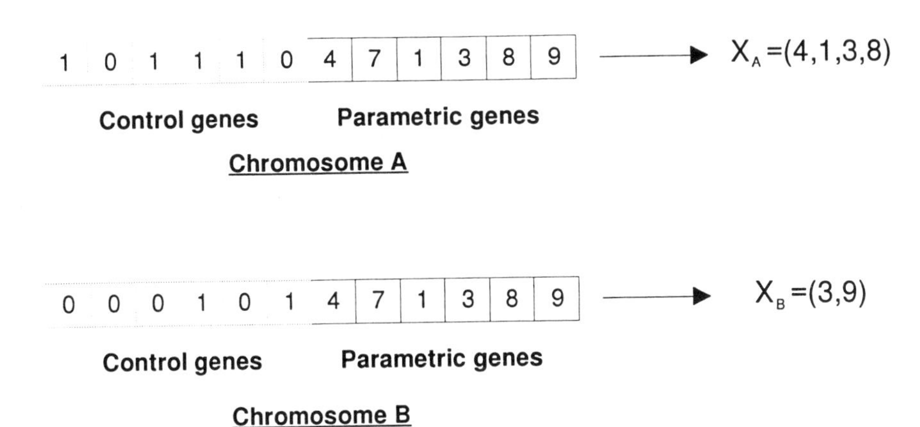

Fig. 5.4. Example of HGA Chromosome Representation

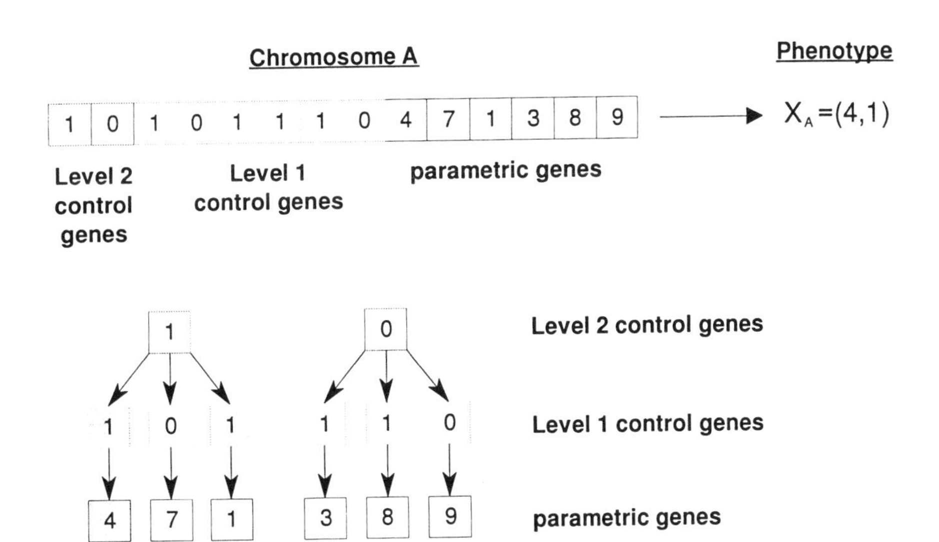

Fig. 5.5. An Example of a 3-level Chromosome

5.3 Genetic Operations

Since the chromosome structure of HGA is fixed, and this is true even for different parameter lengths, there is no extra effort required for reconfiguring the usual genetic operations. Therefore, the standard methods of mutation and crossover may apply independently to each level of genes or even for the whole chromosome if this is homogenous. However, the genetic operations that affect the high-level genes can result in changes within the active genes which eventually leads to a multiple change in the lower level genes. This is the precise reason why the HGA is not only able to obtain a good set of system parameters, but can also reach a minimized system topology.

5.4 Multiple Objective Approach

The basic multiple objective approaches have already been described in Chap. 3. In general, the same approach applies to the HGA. Since its main purpose is to determine the topology of the system, an extra objective function should be included for optimization. Therefore, besides the objective functions $\left(F_i = \begin{bmatrix} f_1 & f_2 & \cdots & f_i \end{bmatrix}^T\right)$ that have been defined by the problem settings as introduced before, another objective (f_{i+1}) is installed for topology optimization.

Based on the specific HGA chromosome structure, the topology information can be acquired from the control genes. Hence, by including the topology information as objective, the problem is now formulated as a multiobjective optimization problem:

$$F = \begin{bmatrix} F_i \\ \\ f_{i+1} \end{bmatrix} \tag{5.1}$$

The GA operation is applied to minimize this objective vector, i.e.

$$\min_{x \in \Phi} F(x) \tag{5.2}$$

where Φ is the searching domain of the chromosome x.

In general, the complexity of the topology can be qualified with integer number, "N". It is assumed that a smaller value of N means a lower order structure which is more desirable.

In order to combine the topological and parametric optimization in a simultaneous manner, let us consider a candidate x_i which is depicted in Fig. 5.6. The candidate x_i has $F_i(x_j) = M_j = \begin{bmatrix} m_1 & m_2 & \cdots & m_i \end{bmatrix} \neq \underline{0}$ and

$f_{i+1}(x_j) = n_j$, and is not a solution for the problem since $\exists k$ s.t. $m_k > 0$. The solution set for the problem is represented by $\{x : F_i(x) = \underline{0}$ and $N_1 \leq f_2(x) \leq N_2\}$. The best solution is denoted by x_{opt} where and $f_{i+1}(x_{opt}) = N_1$. Another solution is to have $f_{i+1}(x_1) = n_1 > N_1$, but, this runs the risk of having a higher order of complexity for the topology.

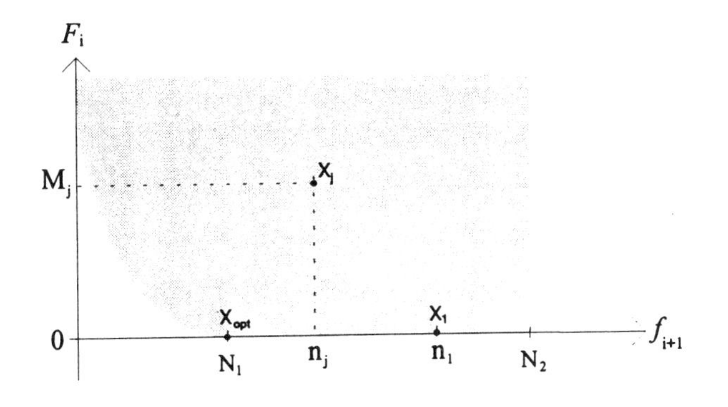

Fig. 5.6. Solution Set for Topology and Parmeteric Optimization Problem

In order to obtain an optimal solution x_{opt}, both (F_i) and $(f_i + 1)$ should therefore be considered simultaneously. Various methods are developed for searching this optimal result.

5.4.1 Iterative Approach

The main difficult in topology optimization is that the degree of complexity of the topology is not known. Therefore, in order to reach an optimal solution, an iterative approach is proposed. The procedure is listed as follows:

1. Let N_2 be the maximum allowable topology for searching. The HGA is applied and terminated when a solution x_1 with $(F_i = \underline{0})$ is obtained. (Fig. 5.7a).
2. Assuming that $f_{i+1}(x_1) = N_3$, the searching domain for the complexity of the topology is reduced from N_2 to $N_3 - 1$. The HGA is then applied again until another solution with $(F_i = \underline{0})$ is obtained. (Fig. 5.7b)
3. Repeat Step 2 by reducing the searching domain of the topology complexity and eventually the optimal point x_{opt} with $f_{i+1}(x_{opt}) = N_1$ will be obtained. (Fig. 5.7c)
4. Another iteration with the complexity of the topology bounded by $N_1 - 1$ is carried out and, of course, no solution may be found. This process can be terminated by setting a maximum number of generations for the HGA. If no solution is found after the generation exceeds this maximum number, the solution obtained in step 3 would be considered as the optimal solution for the problem with lowest complexity. (Fig. 5.7d)

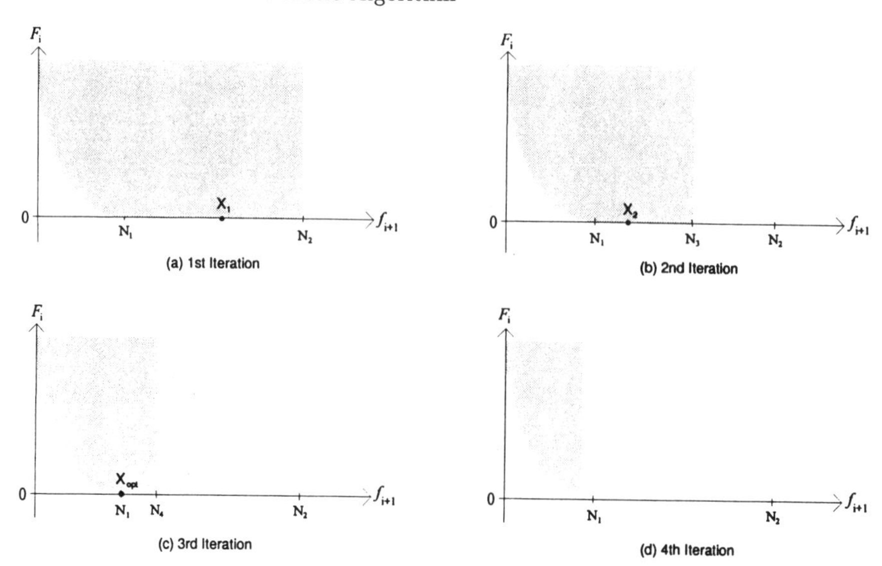

Fig. 5.7. Iterative Approach for Obtaining Optimal Solution with Lowest Complexity

5.4.2 Group Technique

Within a population P, chromosomes can be divided into several sub-groups, G_i, containing the same topology chromosomes (see Fig. 5.8). Each sub-group will contain less than or equal to λ chromosomes, where λ is a pre-defined value, hence,

$$size(G_i) \leq \lambda \quad \forall i = 1, 2, \ldots, M \tag{5.3}$$

The total population size at k-th generation and the maximum population size are expressed in Eqns. 5.4 and 5.5, respectively.

$$P^{(k)} = G_1^{(k)} \cup G_2^{(k)} \cdots \cup G_M^{(k)} \tag{5.4}$$

$$P_{max} = \lambda M \tag{5.5}$$

In order to manage these sub-groups, a modified insertion strategy is developed. The top level description of the insertion strategy for new chromosome z is expressed in Table 5.1.

This grouping method ensures that the chromosomes compete only with those of the same complexity in topological size (f_{i+1}). Therefore, the topology information will not be lost even if the other objective values (F_i) are poor. The selection pressure will still rely on the objective values (F_i) to the problem.

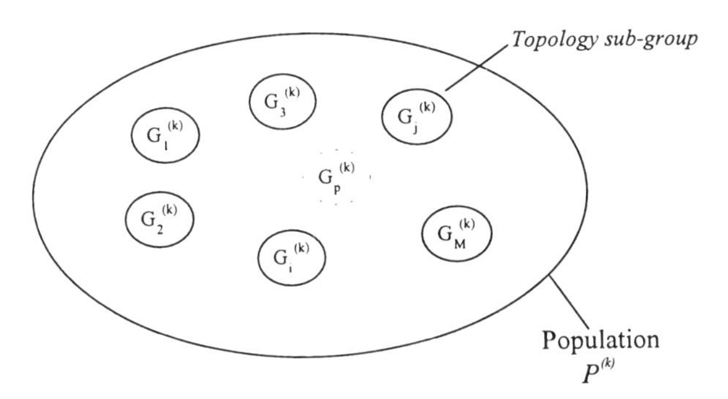

Fig. 5.8. Sub-groups in Population P

Table 5.1. Insertion Strategy

At generation $(k+1)$

Step 1:

If $\left\{G^{(k)}_{f_{i+1}(z)} = \emptyset \quad \text{or} \quad size\left[G^{(k)}_{f_{i+1}(z)}\right] < \lambda\right\}$ then

$\left\{G^{(k+1)}_{f_{i+1}(z)} = G^{(k)}_{f_{i+1}(z)} \cup \{z\} < \lambda\right\}$

else

goto step 2

Step 2:

If $\left\{F_i(z) < F_m = \max\left\{F_i(z_j),\ \forall z_j \in G^{(k)}_{f_{i+1}(z)}\right\}\right\}$ then

$\left\{G^{(k+1)}_{f_{i+1}(z)} = \left\{z_j : F_i(z_j) < F_m,\ z_j \in G^{(k)}_{f_{i+1}(z)}\right\} \cup \{z\}\right\}$

else

goto step 3

Step 3:

Exit

5.4.3 Multiple-Objective Ranking

Instead of using the fitness value for selection, a ranking method involving multiple objective information [40] can be adopted and hence the information of the topology is included.

Chromosome I is then ranked by

$$rank(I) = 1 + p \tag{5.6}$$

if I is dominated by other p chromosomes in the population. The fitness assignment procedure is listed as below:

1. Sort population according to preferable rank;
2. Assign fitness to individual by interpolating from the highest rank to the lowest, and a linear function is used

   $$f(I) = f_{min} + (f_{max} - f_{min})\frac{rank(I) - 1}{N_{ind} - 1} \tag{5.7}$$

   where f_{min} and f_{max} are the lower and upper limits of fitness, respectively; and N_{ind} is the population size.
3. Average the fitness of individual in the same rank, so that all of them will be selected at the same rate.

CHAPTER 6
FILTERING OPTIMIZATION

In system modelling, we often encounter the dilemma of choosing a suitable topological structure to meet the required design criteria. This requirement can sometimes be conflicting, constrained and not always mathematically solvable. Thus far, the black-art technique is still being applied and can be succeeded only by the trial-and-error method. In this chapter, the HGA is demonstrated as being an innovative scheme to tackle problems of this nature. The determination of an Infinite Impulse Response (IIR) filter structure is a classic problem that can be solved by this method. The other application is the optimal low order weighting functions for the design of $\mathbf{H}^{\infty}$ control, using the Loop Shaping Design Procedure (LSDP). Here a detailed account for both design methods is presented, and the essences of applying the HGA as a topological optimizer are accurately described.

6.1 Digital IIR Filter Design

The traditional approach to the design of discrete-time IIR filters involves the transformation of an analogue filter into a digital filter at a given set of prescribed specifications. In general, a bilinear transformation is adopted. Butterworth (BWTH), Chebyshev Type 1 (CHBY1), Chebyshev Type 2 (CHBY2) and Elliptic (ELTC) function based on approximation methods [23, 62, 87, 153] are the most commonly used techniques for implementing the frequency-selective analogue IIR filters. The order of filter is normally determined by the magnitude of the frequency response. The obtained order is considered to be minimal but applicable only to a particular type of filter. This technique for designing the digital filter can be found from the use of MATLAB toolbox [98].

To design the other types of frequency-selective filters such as highpass (HP), bandpass (BP), and bandstop (BS) filters, an algebraic transformation is applied so that the design of HP, BP or BS filters can be derived upon the prototype design of a lowpass (LP) filter [23]. From this transformation, the order of BP and BS filters are found to be twice as much as the LP filter.

However, very often, when we come to the design of a digital IIR filter, an optimal filtering performance is much preferred. This generally consists of the following constraints which are strictly imposed upon the overall design criteria, i.e.

- the determination of lowest filter order;
- the filter must be stable (the poles must lie inside the unit circle); and
- must meet the prescribed tolerance settings.

The first item is not easily determined by any formal method, despite the great amount of effort that has been spent in this area. As for the following two points, these are the constraints which the final filter has to satisfy in order to meet the design performance. These constraints may pose great difficulty for the purpose of optimization in order to meet the design criteria.

Having realized the advantages of HGA formulation, realizing the transformation of the IIR design criteria into HGA presents no problems. Unlike the classic methods, there is no restriction on the type of filter for the design. Each individual filter, whether it is LP, HP, BP or BS, can be independently optimized until the lowest order is reached. This is only possible because of the intrinsic property of the HGA for solving these demanding functions in a simultaneous fashion and, one that involves no an extra cost and effort. To this end, the use of HGA for IIR filter design is summarized as follows:

- the filter can be constructed in any form, such as cascade, parallel, or lattice;
- the LP, HP, BP and BS filters can be independently designed;
- no mapping between analogue-to-digital is required;
- multiobjective functions can be simultaneously solved; and
- the obtained model is of the lowest order.

To illustrate this scheme in detail, Fig. 6.1 depicts the four types of frequency responses that have to be satisfied with various tolerance settings (δ_1, δ_2). The objective of the design is to ensure that these frequency responses are confined within the prescribed frequencies (ω_p, ω_s).

To proceed with the design optimally, the objective functions (f_1, f_2) should be specified in accordance to the specifications of the digital filters. These variables are tabulated in Table 6.1 for each filter structure where $\Delta H_\omega^{(p)}$ and $\Delta H_\omega^{(s)}$ are computed as below:

$$\Delta H_\omega^{(p)} = \begin{cases} \left|H(e^{j\omega})\right| - (1+\delta_1) & \text{if} \quad \left|H(e^{j\omega})\right| > (1+\delta_1) \\ (1-\delta_1) - \left|H(e^{j\omega})\right| & \text{if} \quad \left|H(e^{j\omega})\right| < (1-\delta_1) \end{cases} \tag{6.1}$$

and

$$\Delta H_\omega^{(s)} = \left|H(e^{(j\omega)})\right| - \delta_2 \quad \text{if} \quad \left|H(e^{(j\omega)})\right| > \delta_2 \tag{6.2}$$

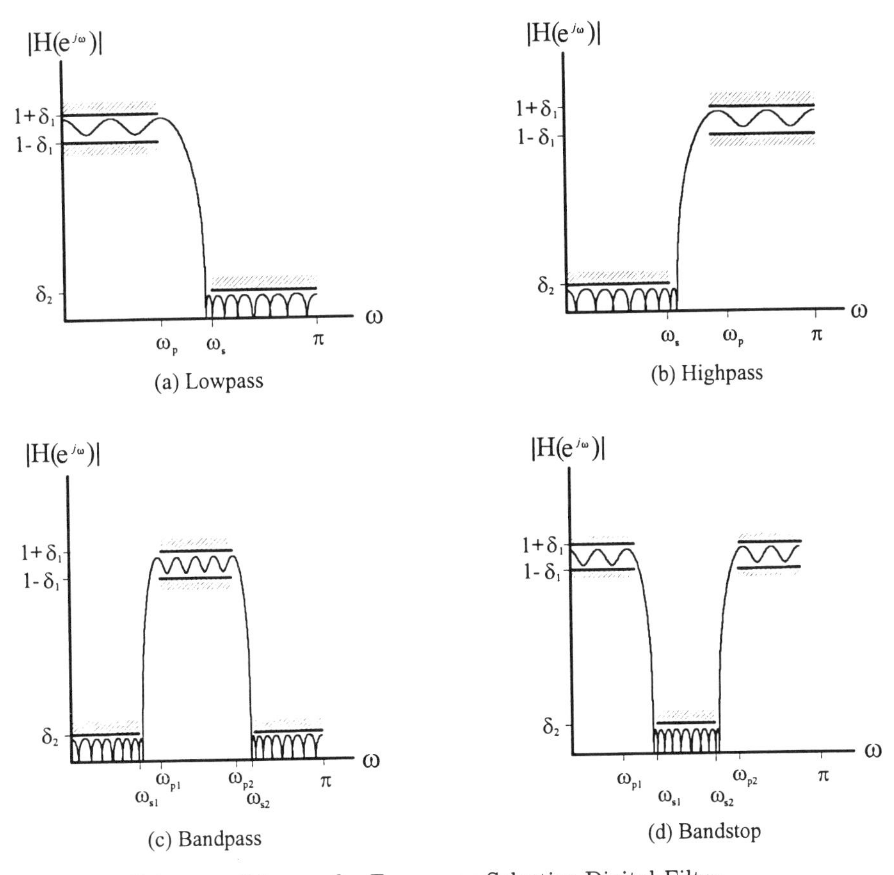

Fig. 6.1. Tolerance Schemes for Frequency-Selective Digital Filter

Table 6.1. Objective Functions

Filter Type	**Objective Functions** $f_1 = \sum \Delta H_\omega^{(p)}$	$f_2 = \sum \Delta H_\omega^{(s)}$
LP	$\forall\omega : 0 \le \omega \le \omega_p$	$\forall\omega : \omega_s \le \omega \le \pi$
HP	$\forall\omega : \omega_p \le \omega \le \pi$	$\forall\omega : 0 \le \omega \le \omega_s$
BP	$\forall\omega : \omega_{p1} \le \omega \le \omega_{p2}$	$\forall\omega : 0 \le \omega \le \omega_{s1}, \quad \omega_{s2} \le \omega \le \pi$
BS	$\forall\omega : 0 \le \omega \le \omega_{p1}, \quad \omega_{p2} \le \omega \le \pi$	$\forall\omega : \omega_{s1} \le \omega \le \omega_{s2}$

6.1.1 Chromosome Coding

The essence of the HGA is its ability to code the system parameters in a hierarchical structure. Despite the format of chromosome organization, this may have to change as, from one system to the other, the arrangement of gene structures can be set in a multilayer fashion. As for the case in IIR filter design, the chromosome representation of a typical pulse transfer function $H(z)$ as indicated in Fig. 6.2 is

$$H(z) = K\frac{(z+b_1)(z+b_2)(z^2+b_{11}z+b_{12})}{(z+a_1)(z+a_2)(z^2+a_{11}z+a_{12})} \tag{6.3}$$

where K is a gain; and a_i, a_{ij}, b_i, b_{ij} $\forall i = j = 1, 2$ are the coefficients.

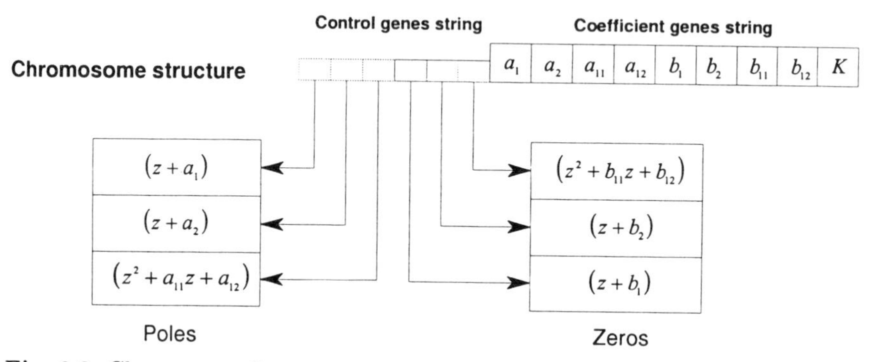

Fig. 6.2. Chromosome Structure

In this configuration, there are two types of gene: one is known as the *control gene* and the other is called the *coefficient gene*. The control gene in bit form decides the state of activation for each block. "1" signifies the state of activation while "0" represents the state of de-activation. The coefficient gene (coded in either binary number or directly represented by a real number) defines the value of the coefficients in each block. The formulation can further be illustrated by an example as follows:

Example 6.1.1. Consider a pulse transfer function:

$$H(z) = 1.5\frac{z+0.1}{z^2+0.7z+0.8} \tag{6.4}$$

The formulation of the genes for a single chromosome, as indicated in Fig. 6.2, is thus

$$[\underbrace{0,0,1,0,1,0}_{Control\ genes},\underbrace{*^{\dagger},*,0.7,0.8,*,0.1,*,*,1.5}_{Coefficient\ genes}] \tag{6.5}$$

In theory, such a chromosome may proceed to the genetic operation and each new generation of chromosomes can be considered as a potential solution to $H(z)$. However, the numerical values of genes may destabilize the filter if their values are not restricted. Since the filter is formed by a combination of first and second order models, the coefficients of the model can be confined within a stable region of an unit circle of z-plane. Hence, the coefficient of the first order model is simply limited to (-1,1), whereas the second order model needs a specific sub-routine to realize such a confinement. In this case, the coefficients of the denominator $(z^2 + a_1 z + a_2)$ must lie within the stability triangle [129] as shown in Fig. 6.3.

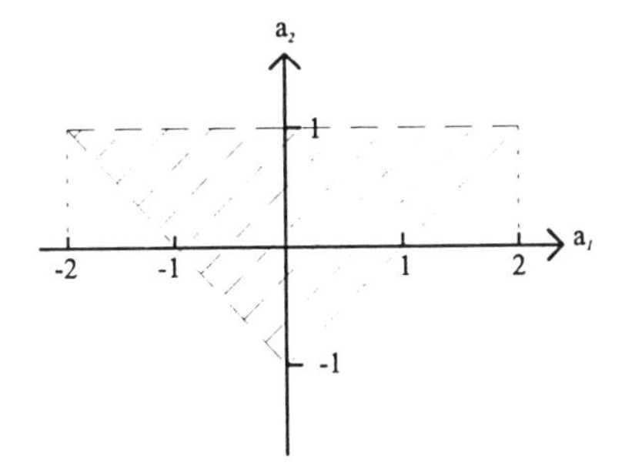

Fig. 6.3. Stability Triangle for $(z^2 + a_1 z + a_2)$

In this way, the parameters of a_2 and a_1 can be initially set into their possible ranges, (-1,1) and (-2,2) respectively. If $(a_1 \leq -1 - a_2)$ or $(a_1 \geq 1 + a_2)$, then a_1 is assigned as a randomly generated number within the range $(-1 - a_2, 1 + a_2)$. This process repeats for each newly generated chromosome, and the condition is checked to ensure that a stable filter is guaranteed. These confinements apply to the constraints of pole location of the filters for both cascade and parallel forms. These are classified and tabulated in Table 6.2.

Table 6.2. Constraints of Pole Locations

	Cascade Form	Parallel Form
Filter $H(z)$	$H(z) = K \prod_{i=1}^{n} \frac{(z+b_i)}{z+a_i} \times \prod_{i=1}^{m} \frac{(z^2+b_{j1}z+b_{j2})}{z^2+a_{j1}z+a_{j2}}$	$H(z) = \sum_{i=1}^{n} K_i \frac{(z+b_i)}{z+a_i} + \sum_{i=1}^{m} K_j \frac{(z^2+b_{j1}z+b_{j2})}{z^2+a_{j1}z+a_{j2}}$
Pole Constraints	$-1 < a_{j2} < 1$ $-1 < a_{j2} < 1 \quad -1 - a_{j2} < a_{j1} < 1 + a_{j2}$	

† Don't care value

6.1.2 The Lowest Filter Order Criterion

Considering that the main purpose of control genes is to classify the topology of the filter structure, therefore, the length of this layer can be set to any length for the representation of a given filter. As this type of gene in the HGA formulation can be automatically turned "on" and "off" while proceeding with the genetic operation, a low order model may be obtained. In this way, those control genes signified "1" imply that an order(s) of the filter has been verified. As a result, the total number of "1"s being ignited would thus determine the final number of orders of the filter.

Based upon this principle, the lowest possible order of the filter may further be reached if a criterion is developed to govern this process. An extra objective function (f_3) has thus been proposed for this purpose. Given that a function f_3 is defined as the order of the filter, then by counting the "1"s within the control genes would be the final filter order of that particular chromosome, that is:

$$f_3 = \sum_{i=1}^{n} p_i + 2\sum_{j=1}^{m} q_j \tag{6.6}$$

where $(n+m)$ is the total length of control genes; p_i and q_j are the control bits governing the activation of i-th first order model and j-th second order model, although the maximum allowable filter order is equal to $(n+2m)$.

In order to ensure f_3 is an optimal value for the filter design, an iterative method which is listed Table 6.3 can be adopted.

Having now formulated the HGA for the filter design, a comprehensive design can be carried out for demonstrating the effectiveness of this approach. The parametric details of the design for the LP, HP, BP and BS filters are listed in Appendix E. For the purpose of comparison, all the genetic operational parameters were set exactly the same for each type of filter. The lowest order of each type of filter was obtained by HGA in this exercise. Table 6.4 summarizes the results obtained by this method and the final filter models can be stated as follows:

$$H_{LP}(z) = 0.0386\frac{(z+0.6884)(z^2-0.0380z+0.8732)}{(z-0.6580)(z^2-1.3628z+0.7122)} \tag{6.7}$$

$$H_{HP}(z) = 0.1807\frac{(z-0.4767)(z^2+0.9036z+0.9136)}{(z+0.3963)(z^2+1.1948z+0.6411)} \tag{6.8}$$

$$H_{BP}(z) = 0.077\frac{(z-0.8846)(z+0.9033)(z^2+0.031z-0.9788)}{(z-0.0592)(z+0.0497)(z^2-0.5505z+0.5371)} \times \frac{(z^2-0.0498z-0.8964)}{(z^2+0.5551z+0.5399)} \tag{6.9}$$

Table 6.3. Operation Procedure

1. Define the basic filter structure used
2. Define the maximum number of iterations, N_{max}
3. Determine the value of n and m for $H(z)$, (see Table 6.2)
4. The optimizing cycle is operated as follows

```
While (tflag ≠ 1)
{
   counter = 0;
   while ((sflag ≠ 1) and (tflag ≠ 1))
   {
      GAcycle(i,j);
      If ( f1 = 0 and f2 = 0 )
      {
         set n,m s.t. (n+2m) = p-1; /*p is the order of the solution*/
         sflag = 1;
      }
      else
         counter++;
      If counter= Nmax
         tflag = 0;
   }
}
```

$$H_{BS}(z) = 0.4919\frac{(z^2 + 0.4230z + 0.9915)(z^2 - 0.4412z + 0.9953)}{(z^2 + 0.5771z + 0.4872)(z^2 - 0.5897z + 0.4838)} \tag{6.10}$$

Table 6.4. Digital Filters by HGA

Filter Type	Lowest Filter Order	Iteration	Objective vs Generation	Transfer Function	Response
LP	3	1487	Fig. 6.5a	H_{LP} (Eqn. 6.7)	Fig. 6.4a
HP	3	1892	Fig. 6.5b	H_{HP} (Eqn. 6.8)	Fig. 6.4b
BP	6	2222	Fig. 6.5c	H_{BP} (Eqn. 6.9)	Fig. 6.4c
BS	4	3365	Fig. 6.5d	H_{BS} (Eqn. 6.10)	Fig. 6.4d

It can seen that the HGA fulfils its promises as regards filter design. Judging from the frequency responses as indicated in Fig. 6.4, all the design criteria have been duly met. In addition, the HGA also provides another capability in which the lowest order for each filter can be independently obtained. This is undoubtedly due to the HGA's ability to solve multiobjective functions in a simultaneous fashion. Such a phenomenon is clearly demonstrated by the performance of objective functions, as indicated in Fig. 6.5, in which a

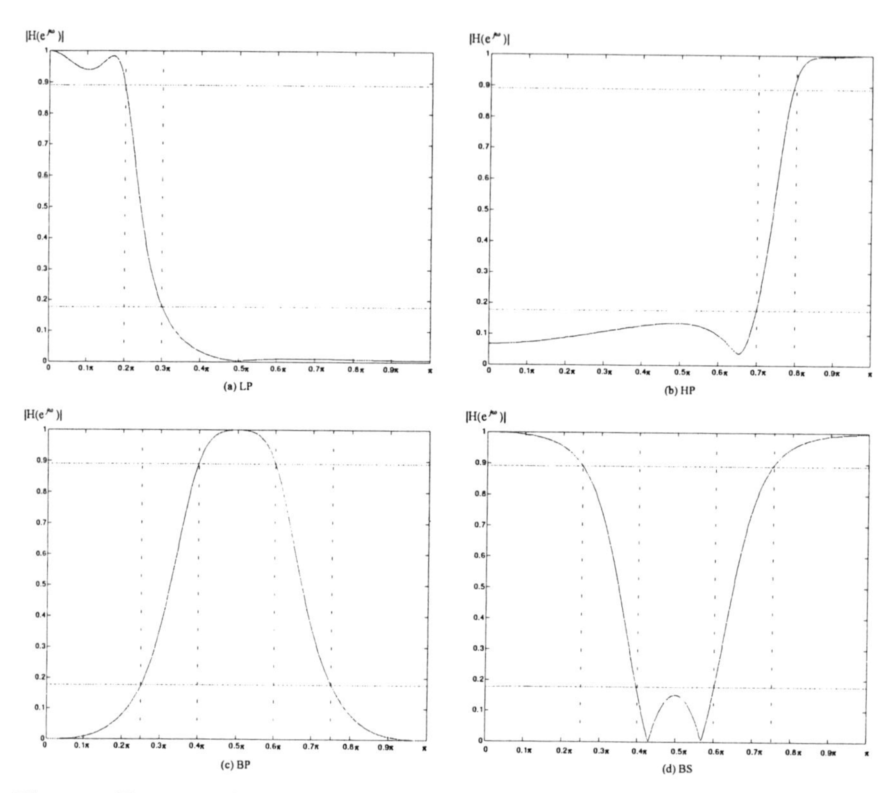

Fig. 6.4. Frequency Response of Filters by HGA

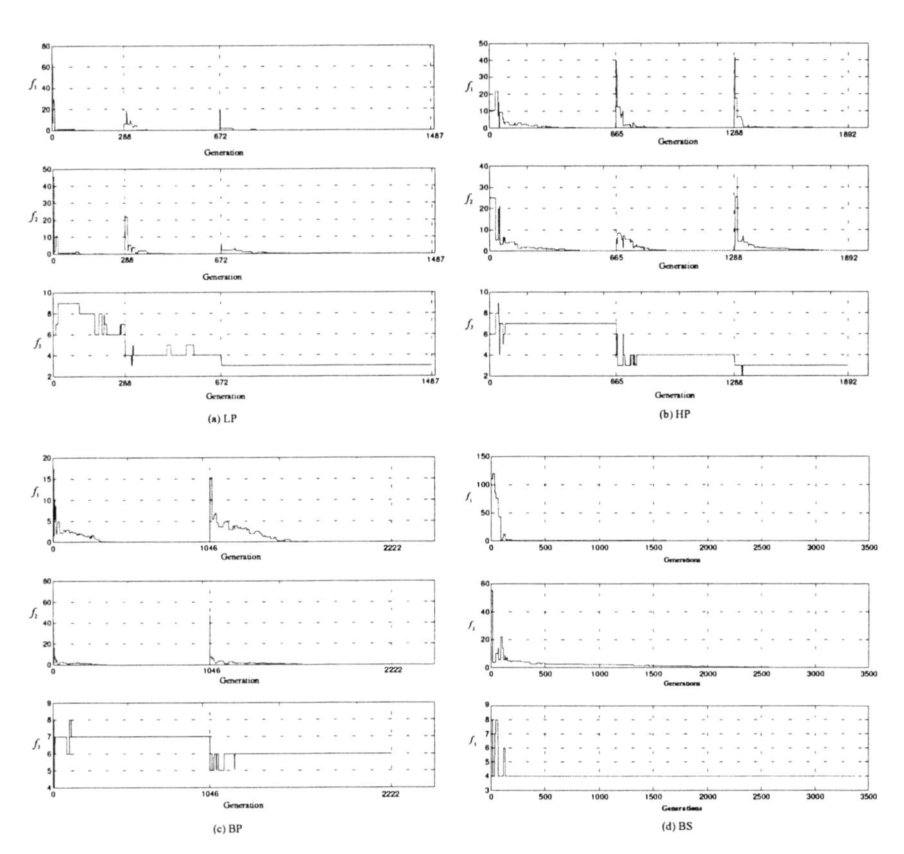

Fig. 6.5. Objective Function vs Generation

lowest order filter is only obtained when the criteria for f_1, f_2 and f_3 are all simultaneously met. However, it should be noted that an adequate filter can be reached even when the functions f_1 and f_2 are fulfilled, while f_3 is not a minimum. This results in a higher order filter which sometimes is acceptable to meet the design specification and greatly shorten the computing effort.

To further highlight the improvement of digital filtering design by the HGA, this method has also been compared with classic filter designs, such as the BWTH, CHBY1, CHBY2 and ELTC approaches. Since the LP filter is the fundamental filter for the design of the other filters, a direct comparison with the LP filter only has been made. The results are summarized and tabulated in Table 6.5. The filter models were found as follows:

$$H_{BWTH} = \frac{0.0007z^6 + 0.0044z^5 + 0.0111z^4 + 0.0148z^3 + \ldots}{z^6 - 3.1836z^5 + 4.6222z^4 - 3.7795z^3 + \ldots} \quad \frac{\ldots\ 0.0111z^2 + 0.0044z + 0.0007}{\ldots\ 1.8136z^2 - 0.4800z + 0.0544} \tag{6.11}$$

$$H_{CHBY1} = \frac{0.0018z^4 + 0.0073z^3 + 0.0110z^2 + 0.0073z + 0.0018}{z^4 - 3.0543z^3 + 3.8290z^2 - 2.2925z + 0.5507} \tag{6.12}$$

$$H_{CHBY2} = \frac{0.1653z^4 - 0.1794z^3 + 0.2848z^2 - 0.1794z + 0.1653}{z^4 - 1.91278z^3 + 1.7263z^2 - 0.6980z + 0.1408} \tag{6.13}$$

$$H_{ELTC} = \frac{0.1214z^3 - 0.0511z^2 - 0.0511z + 0.1214}{z^3 - 2.1112z^2 + 1.7843z - 0.5325} \tag{6.14}$$

It is clearly shown from the frequency responses in Fig. 6.6 that the HGA is much superior to any of the classic methods. Only the ELTC method provides an equivalent order of the filter, but its frequency response is somewhat less attractive. A direct comparison of the magnitude of filter order is tabulated in Table 6.6. It is clearly indicated that the method of HGA for digital filtering design is not only capable of reaching the lowest possible order of the filter, but its ability to satisfy a number of objective functions is also guaranteed. It should also be noted that, the design of HP, BP and BS filters can be independently assigned by the method of HGA, which is a unique asset for designing filters that involve complicated constraints and design requirements.

6.2 H-infinity Controller Design

From the successful experience that we have gained in digital filtering design, the use of the HGA can be expanded into a number of practical engineering applications. The design procedure can be similarly adopted for the solving of control systems design problems that belong to this class. The solution of obtaining an optimal weighting function for $\mathbf{H}^{\infty}$ Loop Shaping Design

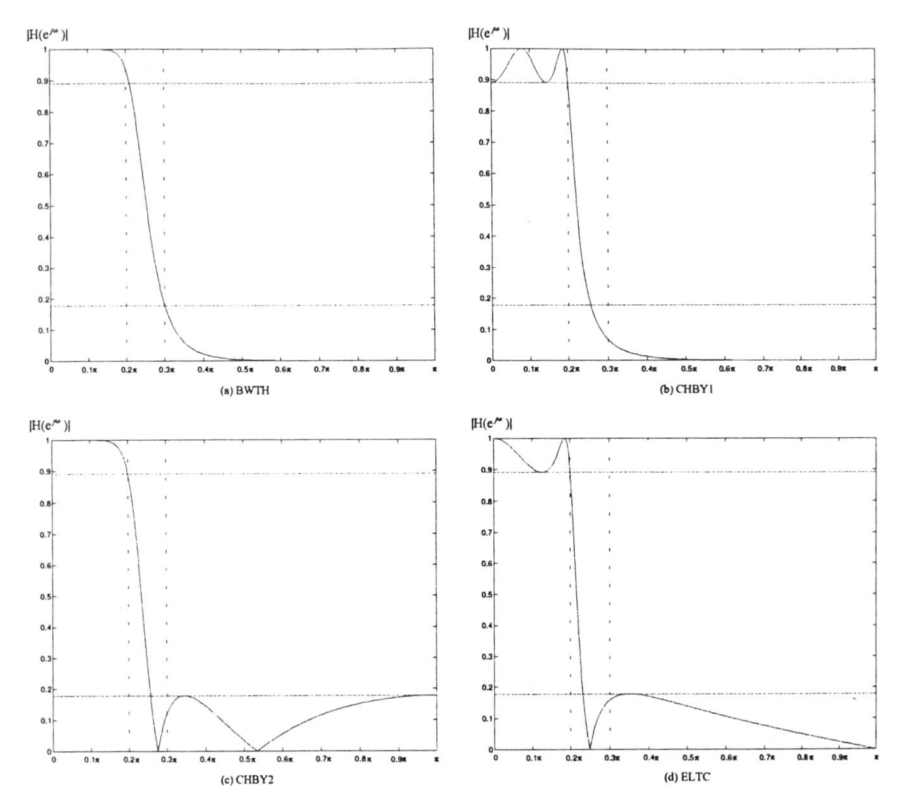

Fig. 6.6. Lowpass Filter Design Using Traditional Methods

Table 6.5. Result of Various Filters Comparison

LP Filter	Lowest Filter Order	Transfer Function	Response
BWTH	6	Eqn. 6.11	Fig. 6.6a
CHBY1	4	Eqn. 6.12	Fig. 6.6b
CHBY2	4	Eqn. 6.13	Fig. 6.6c
ELTC	3	Eqn. 6.14	Fig. 6.6d
HGA	3	Eqn. 6.7	Fig. 6.4a

Table 6.6. Lowest Filter Order due to Various Design Schemes

Filter	BWTH	CHBY1	CHBY2	ELTC	HGA
LP	6	4	4	3	3
HP	6	4	4	3	3
BP	12	8	8	6	6
BS	12	8	8	6	4

Procedure (LSDP) is one of the typical cases for HGA application.

For the last decade or so, $\mathbf{H}^\infty$ optimization has emerged as a powerful tool for robust control system design. This has a sound theoretical background for handling model uncertainties. Based on $\mathbf{H}^\infty$ optimization, a variety of design methods can be developed. The $\mathbf{H}^\infty$ LSDP is one of these that has proven to be effective in practical industrial design. The approach involves the robust stabilization to the additive perturbations of normalized coprime factors of a weighted plant. Prior to robust stabilization, the open-loop singular values are also shaped using weighting functions to give a desired open-loop shape which corresponds to a good closed-loop performance. However, a successful design using LSDP depends on the appropriate choice of weighting functions, which in turn relies on a designer's experience and familiarity with the design approach.

In [154], it has been proposed to enhance the LSDP by combining it with numerical optimization techniques. In order to more effectively search for optimal solutions to the derived constrained optimization problems, the Multiple Objective Genetic Algorithm is suggested in [155]. In this mixed optimization approach, the structures of the weighting functions are to be pre-defined by the designer. It was not possible to search an optimal design systematically among the various structured weights. Therefore, the HGA is a perfect approach to address such a problem.

6.2.1 A Mixed Optimization Design Approach

LSDP is based on the configuration as depicted in Fig.6.7, where $(\tilde{N}, \tilde{M}) \in R\mathbf{H}^\infty$, the space of stable transfer function matrices, is a normalized left coprime factorization of the nominal plant G. That is, $G = \tilde{M}^{-1}\tilde{N}$, and $\exists V, U \in R\mathbf{H}^\infty$ such that $\tilde{M}V + \tilde{N}U = I$, and $\tilde{M}\tilde{M}^* + \tilde{N}\tilde{N}^* = I$; where for a real rational function of s, X^* denotes $X^T(-s)$.

For a minimal realization of $G(s)$

$$G(s) = D + C(sI - A)^{-1}B \stackrel{s}{=} \left[\begin{array}{c|c} A & B \\ \hline C & D \end{array}\right] \tag{6.15}$$

a normalized coprime factorization of G can be given by [99]

$$\left[\begin{array}{cc} \tilde{N} & \tilde{M} \end{array}\right] \stackrel{s}{=} \left[\begin{array}{c|cc} A + HC & B + HD & H \\ \hline R^{-1/2}C & R^{-1/2}D & R^{-1/2} \end{array}\right] \tag{6.16}$$

where $H = -(BD^T + ZC^T)R^{-1}$, $\quad R = I + DD^T$, and the matrix $Z \geq 0$ is the unique stabilizing solution to the algebraic Riccati equation (ARE)

$$\begin{aligned}(A - BS^{-1}D^TC)Z + Z(A - BS^{-1}D^TC)^T& \\ -ZC^TR^{-1}CZ + BS^{-1}B^T = 0&\end{aligned} \tag{6.17}$$

where $S = I + D^T D$

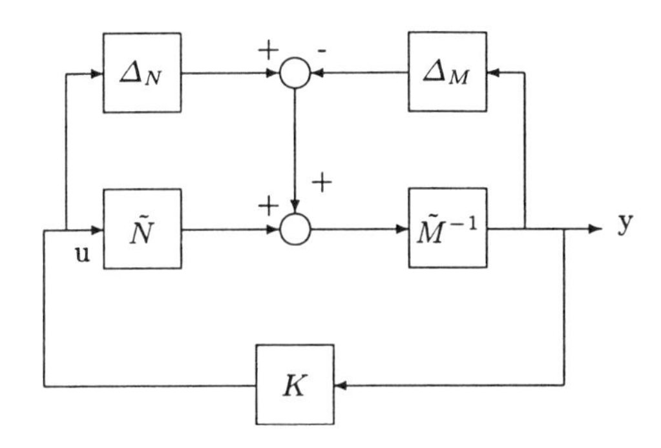

Fig. 6.7. Robust Stabilization with respect to Coprime Factor Uncertainty

A perturbed model G_p of G is defined as

$$G_p = (\tilde{M} + \Delta_M)^{-1}(\tilde{N} + \Delta_N) \tag{6.18}$$

where $\Delta_M, \Delta_N \in R\mathbf{H}^\infty$.

To maximize the class of perturbed models such that the closed-loop system in Fig.6.7 is stabilized by a controller $K(s)$, the $K(s)$ must stabilize the nominal plant G and minimize γ [48] where

$$\gamma = \left\| \begin{bmatrix} K \\ I \end{bmatrix} (I - GK)^{-1}\tilde{M}^{-1} \right\|_\infty \tag{6.19}$$

From the small gain theorem, the closed-loop system will remain stable if

$$\left\| \begin{bmatrix} \Delta_N & \Delta_M \end{bmatrix} \right\|_\infty < \gamma^{-1} \tag{6.20}$$

The minimum value of γ, (γ_0), for all stabilizing controllers is given by

$$\gamma_0 = (1 + \lambda_{max}(ZX))^{1/2} \tag{6.21}$$

where $\lambda_{max}(\cdot)$ represents the maximum eigenvalue, and $X \geq 0$ is the unique stabilizing solution to the following ARE

$$(A - BS^{-1}D^TC)^TX + X(A - BS^{-1}D^TC) \\ -XBS^{-1}B^TX + C^TR^{-1}C = 0 \tag{6.22}$$

A controller which achieves a γ is given in [99] by

$$K \stackrel{s}{=} \left[\begin{array}{c|c} A + BF + \gamma^2(Q^T)^{-1}ZC^T(C + DF) & \gamma^2(Q^T)^{-1}ZC^T \\ \hline B^TX & -D^T \end{array} \right] \tag{6.23}$$

where $F = -S^{-1}(D^TC + B^TX)$ and $Q = (1 - \gamma^2)I + XZ$.

A descriptor system approach may be used to synthesize an optimal controller such that the minimum value γ_0 is achieved.

In practical designs, the plant needs to be weighted to meet closed-loop performance requirements. A design method, known as the LSDP, has been developed [99, 100] to choose the weights by studying the open-loop singular values of the plant, and augmenting the plant with weights so that the weighted plant has an open-loop shape which will give good closed-loop performance.

This loop shaping can be done by the following design procedure:

1. Using a pre-compensator, W_1, and/or a post-compensator, W_2, the singular values of the nominal system G are modified to give a desired loop shape. The nominal system and weighting functions W_1 and W_2 are combined to form the shaped system, G_s, where $G_s = W_2 G W_1$. It is assumed that W_1 and W_2 are such that G_s contains no hidden unstable modes.
2. A feedback controller, K_s, is synthesized which robustly stabilizes the normalized left coprime factorization of G_s, with a stability margin ϵ, and
3. The final feedback controller, K, is then constructed by combining the $\mathbf{H}^\infty$ controller K_s, with the weighting function W_1 and W_2 such that

$$K = W_1 K_s W_2.$$

For a tracking problem, the reference signal is generally fed between K_s and W_1, so that the closed loop transfer function between the reference r and the plant output y becomes

$$Y(s) = (I - G(s)K(s))^{-1} G(s) W_1(s) K_s(0) W_2(0) R(s), \tag{6.24}$$

with the reference r is connected through a gain $K_s(0)W_2(0)$ where

$$K_s(0)W_2(0) = \lim_{s \to 0} K_s(s) W_2(s), \tag{6.25}$$

to ensure unity steady state gain.

The above design procedure can be developed further into a two-degree-of-freedom (2 DOF) scheme as shown in Fig. 6.8.

The philosophy of the 2 DOF scheme is to use the feedback controller $K_s(s)$ to meet the requirements of internal stability, disturbance rejection, measurement noise attenuation, and sensitivity minimization. The pre-compensator K_p is then applied to the reference signal, which optimizes the response of the overall system to the command input. The pre-compensator K_p depends on design objectives and can be synthesized together with the feedback controller in a single step via the $\mathbf{H}^\infty$ LSDP [73].

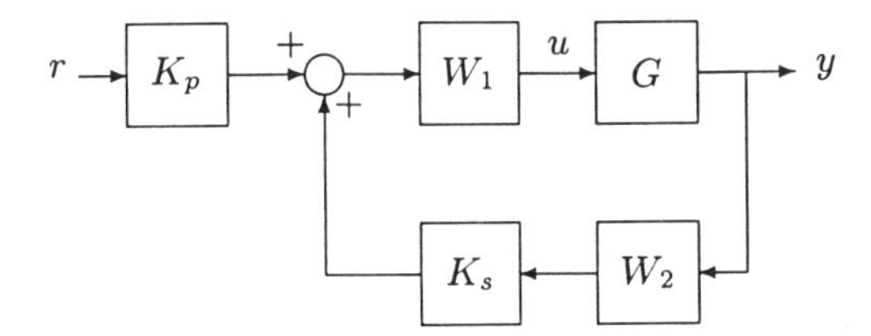

Fig. 6.8. The 2 DOF Scheme

In LSDP the designer has the freedom to choose the weighting functions. Controllers are synthesized directly. The appropriate weighting functions will generate adequate optimal γ_o and will produce a closed-loop system with good robustness and satisfactory and non-conservative performance. The selection of weighting functions is usually done by a trial-and-error method and is based on the designer's experience. In [154], it is proposed to incorporate the Method of Inequalities (MOI) [168] with LSDP such that it is possible to search for "optimal" weighting functions automatically and to meet more explicit design specifications in both the frequency-domain and the time-domain.

In this mixed optimization approach, the weighting functions W_1 and W_2 are the *design parameters.* Control system design specifications are given in a set of inequality constraints. That is, for a given plant $G(s)$, to find (W_1, W_2) such that

$$\gamma_0(G, W_1, W_2) = (1 + \lambda_{max}(ZX))^{1/2} \leq \varepsilon_\gamma \tag{6.26}$$

and

$$\phi_i(G, W_1, W_2) \leq \varepsilon_i; \quad i = 1, 2, \ldots, n \tag{6.27}$$

where ϕ_i's are performance indices, which are algebraic or functional inequalities representing rise-time, overshoot, etc. and ε_γ and ε_i are real numbers representing the desired bounds on γ_0 and ϕ_i, respectively.

Numerical search algorithms may then be applied to find solutions to the above optimization problems.

6.2.2 Hierarchical Genetic Algorithm

The constrained optimization problems derived by the mixed optimization design approach described in Sect. 6.2.1 are usually non-convex, non-smooth and multiobjective with several conflicting design aims which need to be simultaneously achieved. In [155], a Multiobjective Genetic Algorithm [40] is employed to find solutions to such optimization problems. Successful designs have been achieved. It has, however, been found that due to the order and structure of weighting functions being pre-fixed by the designer, the approach lacks flexibility in choosing weights of different forms, which in turn affects the optimality of the design. The HGA has thus been considered to optimize over both the orders and coefficients of the weights, W_1 and W_2, used in the design.

Objective Function and Fitness Function. An objective function f is defined, for a chromosome if Eqn. (6.26) is satisfied, as the number of violated inequalities in Eqn. (6.27). The procedure of objective function evaluation is listed as follows:

1. For a chromosome $I = (W_1, W_2)$ in hierarchy coded form, generate the corresponding W_1 and W_2
2. Calculate $G_s = W_2 G W_1$
3. Find the solutions Z_s, X_s to Eqns. (6.17) and (6.22)
4. Calculate $\gamma_0(W_1, W_2)$ by Eqn. (6.26)
5. Synthesize K_s by Eqn. (6.23)
6. Calculate $\phi_i(G, W_1, W_2)$ for the present chromosome; and
7. Compute f by

$$f = \begin{cases} \sum_{i=1}^{n} m_i & \text{if} \quad \gamma_0 < \varepsilon_\gamma \\ n + 1 + \gamma_0 & \text{else} \end{cases} \tag{6.28}$$

where

$$m_i = \begin{cases} 0 & \text{if} \quad \phi_i \leq \varepsilon_i \\ 1 & \text{else} \end{cases} \tag{6.29}$$

To convert the objective function (f) to the fitness value, a linear ranking approach [158] is applied.

Genetic Operations. Crossover and Mutation on the binary string [27] are applied independently on different levels of a chromosome as in a standard GA.

Optimization Procedure. The optimization procedure is listed as follows:

1. Define the plant G and define the functions ϕ_i;
2. Define the values of ε_i and ε_γ;
3. Define the fundamental form of the weighting functions W_1 and W_2, and the search domain of R_1, R_2;
4. Define the parameters for the HGA;
5. Randomly generate the first population;
6. Calculate the objective value and assign the fitness value to each chromosome;
7. Start the HGA cycle
   - Select parents by the Stochastic Universal Sampling method [7],
   - Generate new chromosomes via Crossover and Mutation,
   - Calculate the objective values of the new chromosomes,

   - Reinsert the new chromosomes into the population and discard the same number of old, low-ranked chromosomes;
8. Terminate if Eqns. (6.26) and (6.27) are satisfied, otherwise repeat the HGA cycle.

6.2.3 The Distillation Column Design

This proposed algorithm has been used to design a feedback control system for the high-purity distillation column described in [88]. The column was considered in its LV* configuration [132], for which the following model was relevant

$$G_D(s,k_1,k_2,\tau_1,\tau_2) = \frac{1}{75s+1}\begin{bmatrix} 0.878 & -0.864 \\ 1.082 & -1.096 \end{bmatrix} \times \begin{bmatrix} k_1 e^{-\tau_1 s} & 0 \\ 0 & k_2 e^{-\tau_2 s} \end{bmatrix} \tag{6.30}$$

where $0.8 \leq k_1, k_2 \leq 1.2$ $0 \leq \tau_1, \tau_2 \leq 1$, and all time units were in minutes.

The time-delay and actuator-gain values used in the nominal model G_n were $k_1 = k_2 = 1$ and $\tau_1 = \tau_2 = 0.5$. The time-delay element was approximated by a first-order Padé approximation for the nominal plant. The design specifications are to design a controller which guarantees for all $0.8 \leq k_1, k_2 \leq 1.2$ and $0 \leq \tau_1, \tau_2 \leq 1$:

1. Closed-loop stability
2. The output response to a step demand $h(t)\begin{bmatrix} 1 \\ 0 \end{bmatrix}$ satisfies $-0.1 \leq y_1(t) \leq 1.1$ for all t, $y_1(t) \geq 0.9$ for all $t > 30$ and $-0.1 \leq y_2(t) \leq 0.5$ for all t.
3. The output response to a step demand $h(t)\begin{bmatrix} 0.4 \\ 0.6 \end{bmatrix}$ satisfies $y_1(t) \leq 0.5$ for all t, $y_1(t) \geq 0.35$ for all $t > 30$ and $y_2(t) \leq 0.7$ for all t, and $y_2(t) \geq 0.55$ for all $t > 30$.
4. The output response to a step demand $h(t)\begin{bmatrix} 0 \\ 1 \end{bmatrix}$ satisfies $-0.1 \leq y_1(t) \leq 0.5$ for all t, $-0.1 \leq y_2(t) \leq 1.1$ for all t and $y_2(t) \geq 0.9$ for all $t > 30$.
5. The frequency response of the closed-loop transfer function between demand input and plant input is gain limited to 50dB and the unity gain crossover frequency of its largest singular value should be less than 150 rad/min.

A set of closed-loop performance functionals $\{\phi_i(G_D, W_1, W_2)$, i=1,2, ..., 16}, are then defined accounting to the design specifications given above. Functionals ϕ_1 to ϕ_{14} are measures of the step response specifications. Functionals ϕ_1, ϕ_6, ϕ_8 and ϕ_{11} are measures of the overshoot; ϕ_4, ϕ_5, ϕ_{13}

* LV indicates that the inputs used are reflux(L) and boilup (V).

and ϕ_{14} are measures of the undershoot; ϕ_2, ϕ_7, ϕ_9 and ϕ_{12} are measures of the rise-time; and ϕ_3 and ϕ_{10} are measures of the cross-coupling. Denoting the output response of the closed-loop system with a plant G_D at a time t to a reference step demand $h(t)\begin{bmatrix} h_1 \\ h_2 \end{bmatrix}$ by $y_i([\ h_1 \quad h_2\ ]^T)$, $i = 1, 2$ the step-response functionals are

$$\phi_1 = \max_t y_1([\ 1 \quad 0\ ]^T, t) \tag{6.31}$$

$$\phi_2 = -\min_{t>30} y_1([\ 1 \quad 0\ ]^T, t), \tag{6.32}$$

$$\phi_3 = \max_t y_2([\ 1 \quad 0\ ]^T, t), \tag{6.33}$$

$$\phi_4 = -\min_t y_1([\ 1 \quad 0\ ]^T, t), \tag{6.34}$$

$$\phi_5 = -\min_t y_2([[\ 1 \quad 0\ ]^T, t), \tag{6.35}$$

$$\phi_6 = \max_t y_1([\ 0.4 \quad 0.6\ ]^T, t), \tag{6.36}$$

$$\phi_7 = -\min_{t>30} y_1([\ 0.4 \quad 0.6\ ]^T, t), \tag{6.37}$$

$$\phi_8 = \max_t y_2([\ 0.4 \quad 0.6\ ]^T, t), \tag{6.38}$$

$$\phi_9 = -\min_{t>30} y_2([\ 0.4 \quad 0.6\ ]^T, t), \tag{6.39}$$

$$\phi_{10} = \max_t y_1([\ 0 \quad 1\ ]^T, t), \tag{6.40}$$

$$\phi_{11} = \max_t y_2([\ 0 \quad 1\ ]^T, t), \tag{6.41}$$

$$\phi_{12} = -\min_{t>30} y_2([\ 0 \quad 1\ ]^T, t) \tag{6.42}$$

$$\phi_{13} = -\min_t y_1([\ 0 \quad 1\ ]^T, t), \tag{6.43}$$

$$\phi_{14} = -\min_t y_2([\ 0 \quad 1\ ]^T, t) \tag{6.44}$$

The steady-state specifications are satisfied automatically by the use of integral action. From the gain requirement in the design specifications, ϕ_{15} is the $\mathbf{H}^\infty$-norm (in DB) of the closed-loop transfer function between the reference and the plant input.

$$\phi_{15} = \sup_\omega \bar{\sigma}\left((I - K(j\omega)G_D(j\omega))^{-1} W_1(j\omega)K_s(0)W_2(0) \right) \tag{6.45}$$

From the bandwidth requirement in the design specification, ϕ_{16} is defined (in rad/min) as

$$\phi_{16} = \max\{\omega\} \quad \text{such that}$$
$$\bar{\sigma}\left((I - K(j\omega)G_D(j\omega))^{-1} W_1(j\omega)K_s(0)W_2(0) \right) \geq 1 \tag{6.46}$$

The fundamental structures of W_1 and W_2 in the design examples are given as:

$$W_1 = \frac{(s+w_5)(s+w_6)(s^2+w_7 s+w_8)}{s(s+w_1)(s+w_2)(s^2+w_3 s+w_4)} \begin{bmatrix} \alpha_1 & 0 \\ 0 & \alpha_2 \end{bmatrix} \quad (6.47)$$

$$W_2 = \frac{(s+w_{13})(s+w_{14})(s^2+w_{15} s+w_{16})}{(s+w_9)(s+w_{10})(s^2+w_{11} s+w_{12})} \begin{bmatrix} \alpha_3 & 0 \\ 0 & \alpha_4 \end{bmatrix} \quad (6.48)$$

In general, W_1 and W_2 can be diagonal matrices with different diagonal elements.

The chromosome is a binary string describing the control and coefficient genes, g_c and g_r where

$$g_c \in B^{12}$$
$$g_r = \{w_1, w_2, \ldots, w_{16}, \alpha_1, \alpha_2, \alpha_3, \alpha_4\} \in R_1^{16} \times R_2^4$$

where $B = [0, 1]$ and R_1, R_2 defining the search domain for the parameters, which usually represents an admissible region, e.g. ensuring that the weighting functions are stable and minimum phase.

Case Study A: Optimization of Nominal Plant Specifications with Time Delay $\tau_1 = \tau_2 = 0.5$

This proposed algorithm has been used to satisfy the performance design specification for the nominal plant G_n using the configuration of Fig. 6.8. The design criteria are derived from Eqns. (6.26) and (6.27)

$$\gamma_0(G_n, W_1, W_2) \leq \varepsilon_\gamma \quad (6.49)$$
$$\phi_i(G_n, W_1, W_2) \leq \varepsilon_i \quad \text{for } i = 1, 2, \ldots, 16 \quad (6.50)$$

For stability robustness, the value of ε_γ should not be too large, and is here taken as

$$\varepsilon_\gamma = 5.0 \quad (6.51)$$

The performance functionals $\phi_i(G_n, W_1, W_2)$ and the respective prescribed bounds are decided from the design specifications and are shown in second column of Table 6.9. The parameters of the HGA used in the simulation are tabulated in Table 6.7.

It took about 135 generations to obtain the optimal compensators. The weighting functions obtained were:

$$W_1 = \frac{(s+1.2800)(s+1.5005)}{s(s+0.8215)(s+1.4868)} \begin{bmatrix} 2.4390 & 0 \\ 0 & 5.8533 \end{bmatrix}$$

$$W_2 = \frac{(s+1.7873)(s^2+0.5620s+1.9844)}{(s+1.7385)(s^2+1.4946s+1.8517)} \begin{bmatrix} 36.0976 & 0 \\ 0 & 36.5854 \end{bmatrix}$$

with $\gamma_0 = 3.6147$ which successfully satisfy Eqns. (6.49) and (6.50). The convergence of the objective value is plotted in Fig. 6.9.

† Number of New Chromosomes Generated = Generation Gap × Population Size

Table 6.7. Parameter Setting of HGA

Population Size		40
Generation Gap†		0.2
	Control Gene	
Resolution		1 bit
Range	B	[0, 1]
Crossover	1-point Crossover	Crossover Rate = 0.7
Mutation	Bit Mutation	Mutation Rate = 0.05
	Coefficient Gene	
Resolution		10 bits
Range	R_1	(0,2)
	R_2	(0,500)
Crossover	3-point Crossover	Crossover Rate = 0.8
Mutation	Bit mutation	Mutation Rate = 0.1

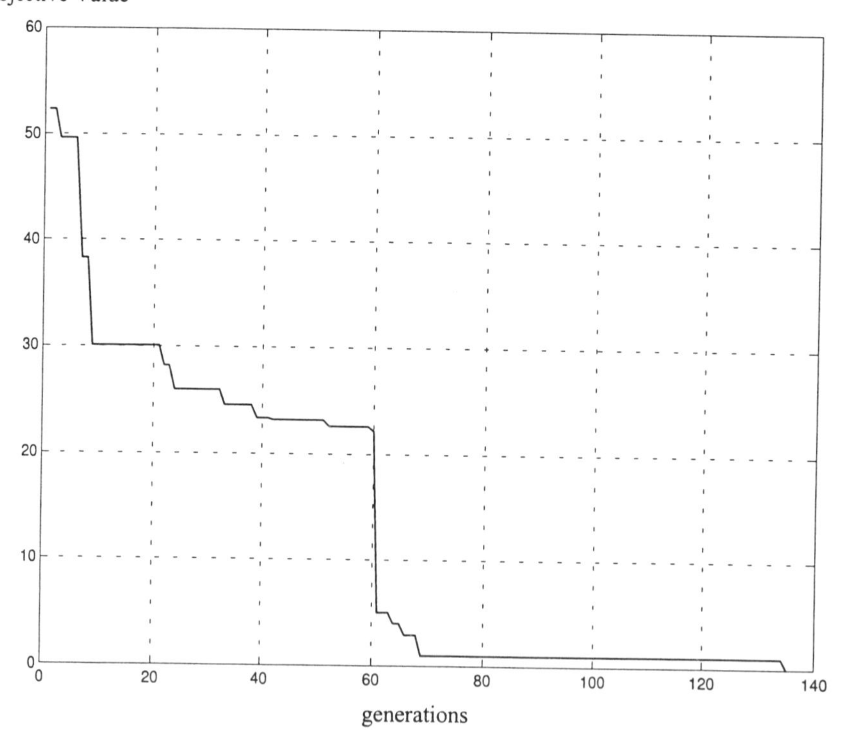

Fig. 6.9. Objective Value vs Generations

Extreme plants G_1, G_2, G_3, G_4 with system parameters shown in Table 6.8 were used for testing the system's robustness. These extreme plant models were judged to be the most difficult to obtain simultaneously good performance and it was found that the final system was not very robust.

Table 6.8. Extreme Plants G_j for $j = 1, 2, 3, 4$

	τ_1	τ_2	k_1	k_2
G_1	0	0	0.8	0.8
G_2	1	1	0.8	1.2
G_3	1	1	1.2	0.8
G_4	1	1	1.2	1.2

Case Study B: Optimization of Plant Specifications with Time Delay $\tau_1 = \tau_2 = 1$

The setting is the same as that for *Case study A*, except that the design criteria, Eqn. (6.50), are modified. This is to realize the design criteria when $\tau_1 = \tau_2 = 1$ which are considered to be more difficult to achieve. The design criteria are modified as follows:

$$\begin{aligned} &\gamma_0(G_n, W_1, W_2) \leq \varepsilon_\gamma \\ &\phi_i(G_m, W_1, W_2) \leq \varepsilon_i \quad for\ i = 1, 2, \ldots, 16 \end{aligned} \tag{6.52}$$

where G_m is the plant with $k_1 = k_2 = 1$ and $\tau_1 = \tau_2 = 1$ using a fifth-order Padé approximation.

It took about 800 generations to obtain W_1 and W_2 for robust feedback control. The parameters of W_1 and W_2 were

$$\begin{aligned} W_1 &= \frac{(s+0.8878)(s^2+0.3161s+1.1044)}{s(s+1.1278)(s^2+1.1356s+0.1444)} \begin{bmatrix} 13.1707 & 0 \\ 0 & 13.1707 \end{bmatrix} \\ W_2 &= \frac{(s+0.1678)(s^2+0.3727s+0.7161)}{(s+1.5727)(s^2+1.0049s+1.8946)} \begin{bmatrix} 50.2439 & 0 \\ 0 & 52.1951 \end{bmatrix} \end{aligned}$$

where $\gamma_0 = 3.3047$.

The closed-loop performances are tabulated in Table 6.9 and depicted in Fig. 6.10. All the design criteria are satisfied except that the 50dB gain limit is marginally exceeded by $\phi_{15}(G_2)$ and $\phi_{15}(G_3)$.

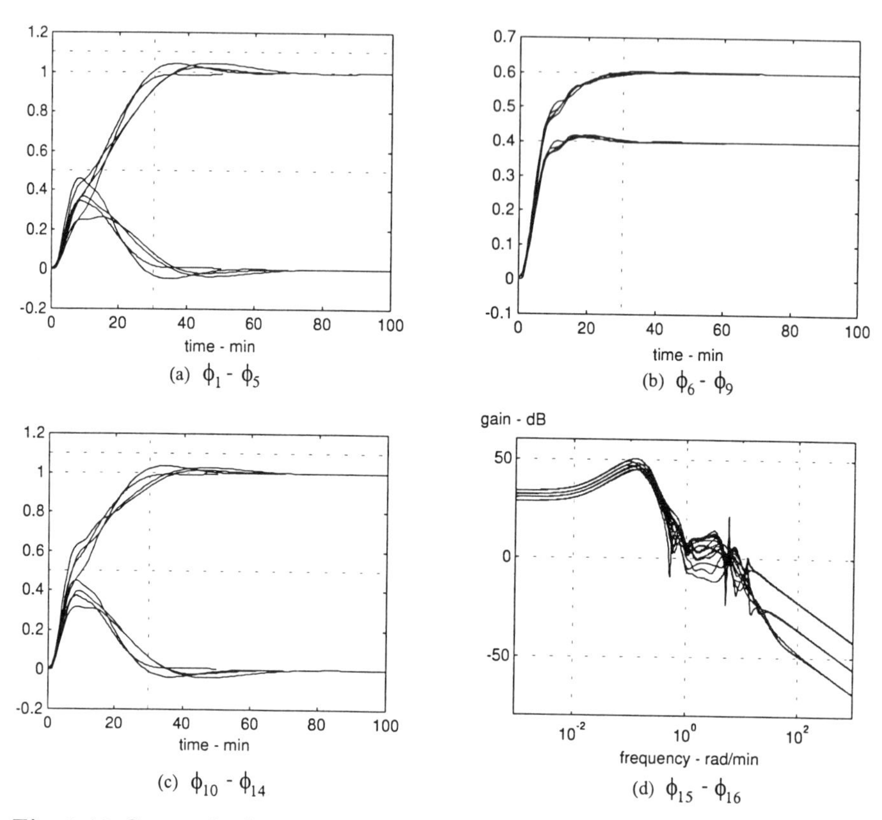

Fig. 6.10. System Performance for Optimization of Plant Specifications with Time Delay $\tau_1 = \tau_2 = 1$

Table 6.9. Final System Performance for Optimization of Plant Specifications with Time Delay $\tau_1 = \tau_2 = 1$

i	ε_i	$\phi_i(G_m)$	$\phi_i(G_1)$	$\phi_i(G_2)$	$\phi_i(G_3)$	$\phi_i(G_4)$
1	1.1	1.0212	1.0410	1.0377	1.0306	1.0030
2	-0.9	-0.9732	-0.9655	-0.9898	-0.9870	-0.9687
3	0.5	0.3556	0.3625	0.2844	0.4290	0.3471
4	0.1	0	0	0.0001	0.0002	0
5	0.1	0.0159	0.0305	0.0232	0.0477	0.0054
6	0.5	0.4177	0.4198	0.4272	0.4274	0.4308
7	-0.35	-0.3981	-0.3966	-0.3851	-0.3873	-0.3843
8	0.7	0.6012	0.6023	0.6149	0.6162	0.6191
9	-0.55	-0.5972	-0.5967	-0.5838	-0.5873	-0.5777
10	0.5	0.3837	0.3912	0.4300	0.3359	0.3746
11	1.1	1.0124	1.0242	1.0299	1.0372	1.0138
12	-0.9	-0.9832	-0.9786	-0.9748	-0.9841	-0.9660
13	0.1	0.0171	0.0330	0.0417	0.0270	0.0058
14	0.1	0	0	0.0003	0.0002	0
15	50.0	48.4479	49.7050	51.0427	51.0167	47.76
16	150.0	9.0773	10.7159	9.8627	10.000	9.8627

Case Study C: Optimization of Overall Plants Specifications with Extreme Conditions

Since it may not be easy to obtain a controller satisfying the performance specifications for those extreme plant models by optimization of the nominal plant or a typical plant, an alternative will simultaneously optimize all the extreme plants. The design criteria are now re-defined as

$$\gamma_0(W_1, W_2) \leq \varepsilon_\gamma \tag{6.53}$$

$$\phi_i(G_j, W_1, W_2) \leq \varepsilon_i \quad \text{for } i = 1, 2, \ldots, 16;\ j = 1, 2, 3, 4 \tag{6.54}$$

A Multiple Objective HGA (MO-HGA) has been applied here. Multiple objective ranking [40] approach is used. The chromosome I is ranked as

$$rank(I) = 1 + p \tag{6.55}$$

if I is dominated by other p chromosomes in the population.

From Eqn. (6.54), 64 objectives need to be achieved. Such huge numbers of objectives demand a large number of comparison operations. Hence, these have been simplified into 4 objectives to indicate the fitness for each extreme plant as before. Define m_{ij} for extreme plant $i = 1, 2, \ldots, 16$ and $j = 1, 2, 3, 4$ as

$$m_{ij} = \begin{cases} 0 & \text{if} \quad \phi_i(G_j, W_1, W_2) \leq \varepsilon_i \\ 1 & \text{else} \end{cases} \tag{6.56}$$

The objective f_j for extreme plant j, for $j = 1, 2, 3, 4$, is

$$f_j = \begin{cases} \sum_{i=1}^{n} m_{ij} & \text{if} \quad \gamma_0 < \varepsilon_\gamma \\ n + 1 + \gamma_0 & \text{else} \end{cases} \tag{6.57}$$

where $n = 16$.

After 448 generations, W_1 and W_2 were obtained and expressed as follows:

$$W_1 = \frac{(s+0.8956)(s^2+0.7161s+1.4888)}{s(s+1.7249)(s^2+1.9122s+0.1444)} \begin{bmatrix} 76.0976 & 0 \\ 0 & 47.3171 \end{bmatrix}$$
$$W_2 = \frac{(s+1.4537)(s^2+0.1990s+0.5444)}{(s+1.4498)(s^2+1.2741s+1.9551)} \begin{bmatrix} 17.0732 & 0 \\ 0 & 17.5610 \end{bmatrix}$$

with $\gamma_0 = 3.1778$. Fig. 6.11 demonstrates the multiple objective optimization process of the proposed MO-HGA. Trade-offs between different objective values can be noticed.

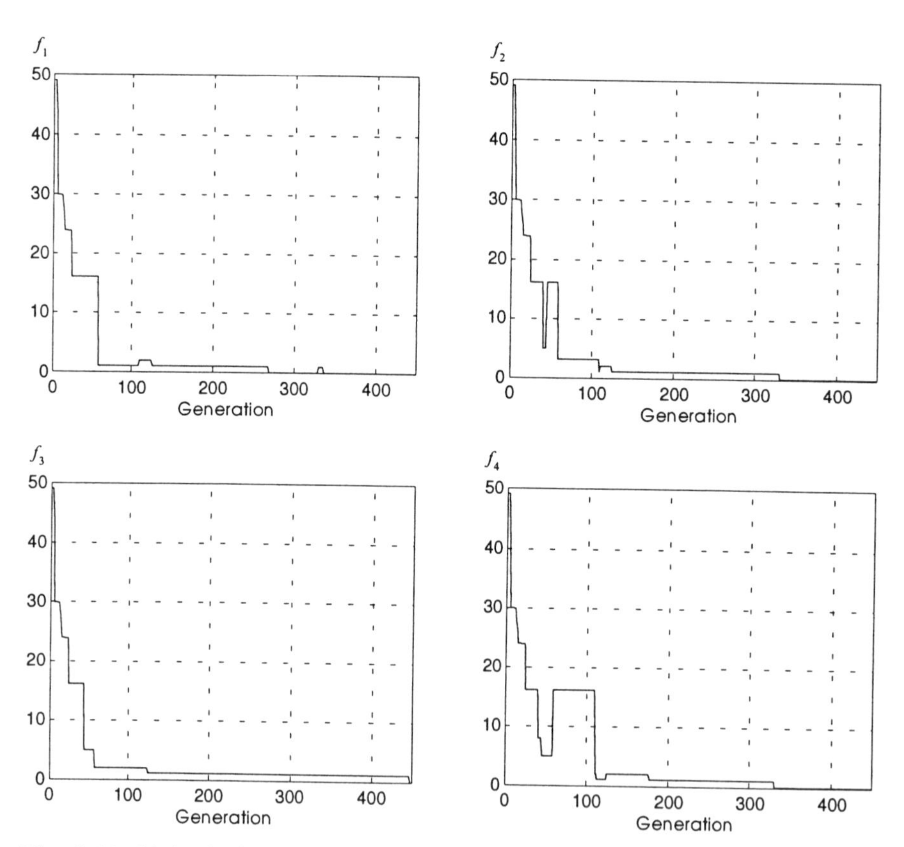

Fig. 6.11. Multiple Objective Values vs Generations

The closed loop system responses for the extreme plants are tabulated in Table 6.10 and depicted in Fig. 6.12.

Table 6.10. Final System Performance for Optimization of Overall Plants Specifications with Extreme Conditions

i	ε_i	$\phi_i(G_1)$	$\phi_i(G_2)$	$\phi_i(G_3)$	$\phi_i(G_4)$
1	1.1	1.0477	1.0142	1.0401	1.0024
2	-0.9	-0.9159	-0.9791	-0.9125	-0.9525
3	0.5	0.3843	0.3204	0.4193	0.3578
4	0.1	0	0.0001	0.0001	0.0001
5	0.1	0.0363	0.0168	0.0392	0.0018
6	0.5	0.4266	0.4242	0.4251	0.4218
7	-0.35	-0.3960	-0.3967	-0.3963	-0.3998
8	0.7	0.6030	0.6023	0.6034	0.6003
9	-0.55	-0.5938	-0.5915	-0.5976	-0.5969
10	0.5	0.4065	0.4604	0.3569	0.3786
11	1.1	1.0292	1.0128	1.0315	1.0015
12	-0.9	-0.9478	-0.9409	-0.9921	-0.9709
13	0.1	0.0384	0.0139	0.0328	0.0019
14	0.1	0	0.0002	0.0001	0.0002
15	50.0	48.8978	49.5575	49.7752	45.8310
16	150.0	16.9133	13.1862	11.6430	11.9696

6.2.4 Design Comments

The proposed HGA enables the simultaneous searching on the structures and coefficients of the weighting functions. It is an unique approach for such a design. Several advantages have been gained from this method:

- The HGA can easily handle the constraints to ensure the stability of the weight functions;
- a multiple objective approach can be adopted to address the conflicting control design specifications; and
- the structures of the weighting functions are no longer pre-fixed; but only a fundamental structure is required, which provides the optimality for the solution over several different types of weights.

In the case studies, the performance was evaluated for a selection of extreme plant models chosen by the designer. The problem of efficiently determining the worst-case performance over the range of plants still remains. Since the proposed algorithm follows the formulation of an MOI which requires the choice of several plants only, it is necessary to choose the most representative plants out of all possible plant models.

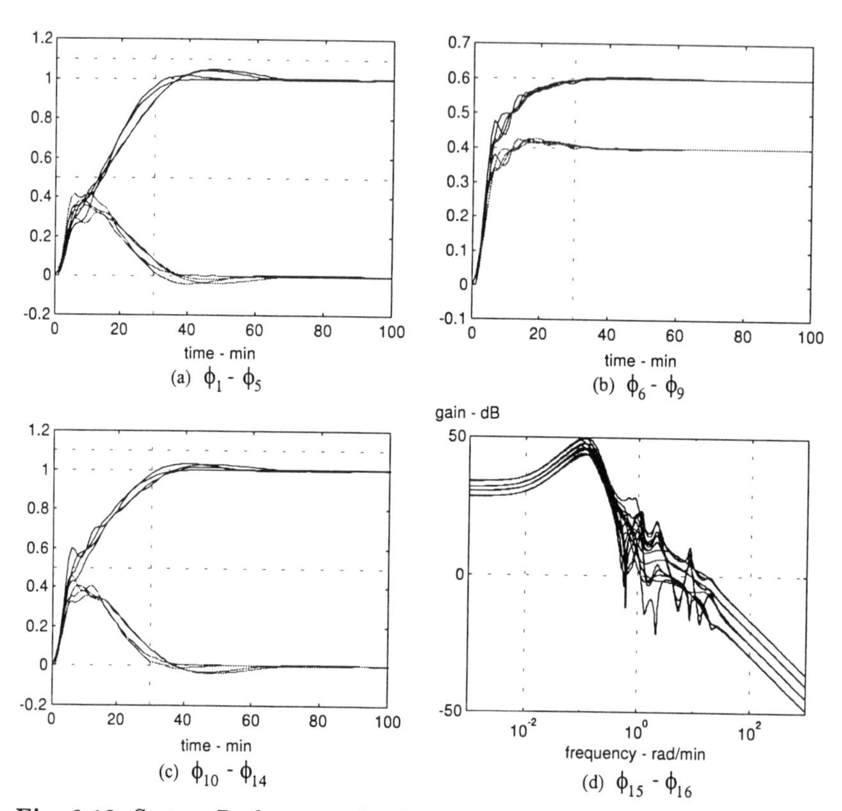

Fig. 6.12. System Performance for Optimization of Overall Plants Specifications with Extreme Conditions

CHAPTER 7
EMERGING TECHNOLOGY

It is anticipated that future engineering design will disregard its own discipline and will become heavily involved with artificial intelligence (AI). This trend of development is understandable since computing power has become so much faster and cheaper nowadays that a required solution can be automatically obtained even when it is based upon a computationally intensive scheme.

The use of AI is widespread and forms a core unit for the production of this emerging technology that is being utilized in both academic and industrial domains. The noticeable development of neural networks and fuzzy logic systems for engineering applications are just two typical examples that illustrate this point of view. Very often, when AI is applied, and despite the brilliance of the technology, a prior knowledge of the system's topology and its governing parameters of concern is required in order to fully exploit the use of AI. Unfortunately, this is not easily come by and sometimes can only be achieved by sheer computing power.

Having discovered the attributes of the HGA, in Chap. 6, for solving the topological structure of filtering problems, in this chapter, the HGA is also used to solve problems of a similar nature. This time, the HGA is used to tackle the well known neural network (NN) topology as well as the fuzzy logic membership functions and rule problems. It is believed from the investigations that the HGA offers a sound approach to reach an optimal but reduced topology in both cases. It is our belief that the HGA will become a potent technology, and that this trend should be encouraged for future system design.

7.1 Neural Networks

The use of NN for system control and signal processing has been well accepted. The most noticeable applications are in the areas of telecommunication, active noise control, pattern recognition, prediction and financial analysis, process control, speech recognition, etc. [4, 161]. The widespread use of NN is due mainly to its behavioural emulation of the nature of the human brain and the fact that its structure can be mathematically formulated. An efficient NN topology is capable of enhancing the system performance in terms of learning speed, accuracy, noise resistance and generalization ability. Hence, NN can be considered as a parallel and distributed processing unit that consists of multiple processing elements. This structure is complex, unknown and must be pre-defined initially.

To this end, the required technique to obtain an optimal NN topology can be generalized into:

1. a prior analysis of the potentialities of a network;
2. optimum size of a network that reduces the enormous search spaces in learning, (to improve the computational power);
3. use of a computationally prohibitive construction-destruction algorithm [16, 108]; and
4. use of mathematical methods to determine the architecture and parameters of the network [166].

The recent development of AI to explore a different approach to optimize the network configuration has caused great interest in the NN area. The application of fuzzy logic techniques to adapt the network configuration [128] has been reported. Evolutionary Algorithms have been applied to determine the network construction and/or the connection weight optimization [3, 102, 103]. The promising results obtained by using GA [97, 112] to train and design artificial neural networks has proved to be a useful technique for integration. All these techniques provide viable alternatives for the improvement of NN topology.

In this section, the HGA is proposed for the optimization of NN topology. The advantage of this approach is that the genes of the chromosome are classified into two categories in a hierarchical form. This is an ideal formulation for the genes as the layers, neurons, connection weightings and bias can be formed within the string of chromosomes. This provides a vast dimension for genetic operation which, in the end, improves the computational power and the optimization of NN topology.

7.1.1 Introduction of Neural Network

A general neural network topology takes the form of a multilayer feedforward network as depicted in Fig. 7.1.

The basic processing element is called a neuron. A neuron consists of an activity level, a set of input and output connections with a bias value associated to each connection. Fig. 7.2 shows a typical neuron with n-input connections and a single output connection.

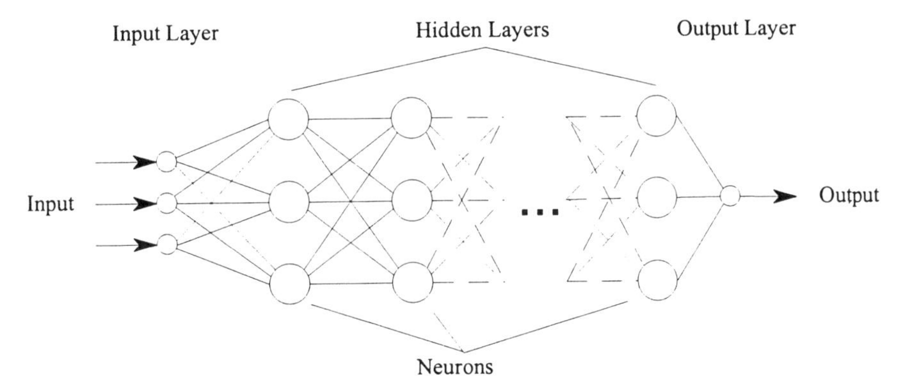

Fig. 7.1. A Multilayer Feedforward Neural Network Topology

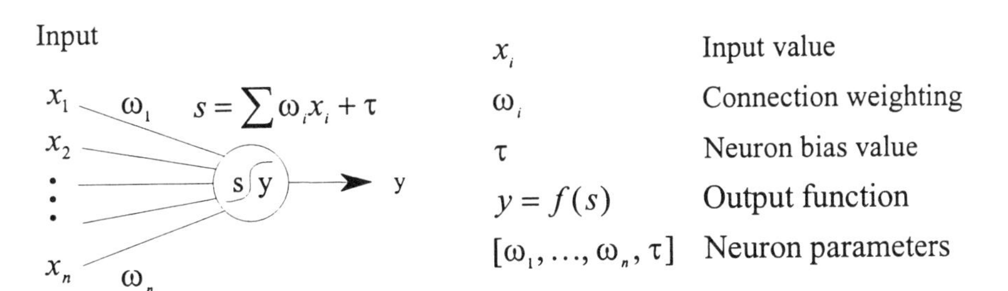

Fig. 7.2. A Single Neuron

The output of the neuron is determined as:

$$y = f\left(\sum_{i=1}^{n} \omega_i x_i + \tau\right) \tag{7.1}$$

where $x_1, x_2, \ldots, x_n$ are input signals; $\omega_1, \omega_2, \ldots, \omega_n$ are connection weightings; τ is the bias value; and f is a defined output function that may be a sigmoid, tanh, step function etc.

In order for this topology to function according to design criteria, a learning algorithm that is capable of modifying the network parameters, as indicated in Eqn. 7.1, is of paramount important to the NN. The backpropagation (BP) technique employs a gradient descent learning algorithm [130] that is commonly used by the NN community. This approach suffers from a pre-defined topology such that the numbers of neurons and connections must be known a prior. Furthermore, as the network complexity increases, the performance of BP decreases rapidly. The other deficit of BP is its use of gradient search algorithms, where discontinuous connection weightings cannot be handled.

7.1.2 HGA Trained Neural Network (HGANN)

Having realized the pros and cons of NN, the bottle-neck problem lies within the optimization procedures that are implemented to obtain an optimal NN topology. Hence, the formulation of the HGA is applied for this purpose [145]. The HGA differs from the standard GA with a hierarchy structure in that each chromosome consists of multilevels of genes.

Fig. 7.3 shows the chromosome representation in the HGANN system. Each chromosome consists of two types of genes, i.e. control genes and connection genes. The control genes in the form of bits, are the genes for layers and neurons for activation. The connection genes, a real- value representation, are the genes for connection weightings and neuron bias. A neural network defined by this structure is depicted in Fig. 7.3.

Within such a specific treatment, a structural chromosome incorporates both active and inactive genes. It should be noted that the inactive genes remain in the chromosome structure and can be carried forward for further generations. Such an inherent genetic variation in the chromosome avoids any trapping at local optima which has the potential to cause premature convergence. Thus it maintains a balance between exploiting its accumulated knowledge and exploring the new areas of the search space. This structure also allows larger genetic variations in chromosome while maintaining high viability by permitting multiple simultaneous genetic changes. As a result, a single change in high level genes will cause multiple changes (activation

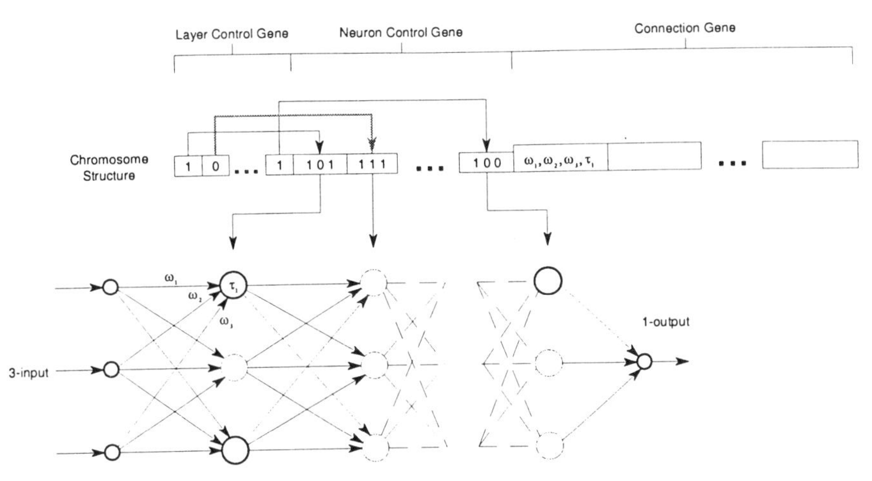

Fig. 7.3. HGANN Chromosome Structure

or deactivation in the whole level) in lower level genes. In the case of the traditional GA, this is only possible when a sequence of many random changes takes place. Hence the computational power is greatly improved.

To formulate such an HGANN, its overall system block diagram is shown in Fig. 7.4. This hierarchical genetic trained neural network structure has the ability to learn the network topology and the associated weighting connection concurrently. Each learning cycle is known as a generation.

Population. In order to explore all possible topologies, the population of HGANN at k-th generation, $P^{(k)}$, is divided into several connection sub-groups,

$$\begin{aligned} G_1^{(k)} \cup G_2^{(k)} \ldots \cup G_M^{(k)} &= P^{(k)} \quad \text{and} \\ G_i^{(k)} \cap G_j^{(k)} &= \phi, \quad \forall i \neq j \end{aligned} \tag{7.2}$$

where M is the maximum number of possible connections represented by HGANN; and $G_i^{(k)}$ is the subgroup of chromosomes that represents those networks with i active connection at k-th generation.

A concurrent factor, λ, is used to define the maximum number of chromosome stored in each sub-group,

$$size\left[G_i^{(k)}\right] \leq \lambda \tag{7.3}$$

where $size\left[G_i^{(k)}\right]$ is the number of elements in $G_i^{(k)}$.

Hence, the maximum population size is limited to P_{max} which is defined as

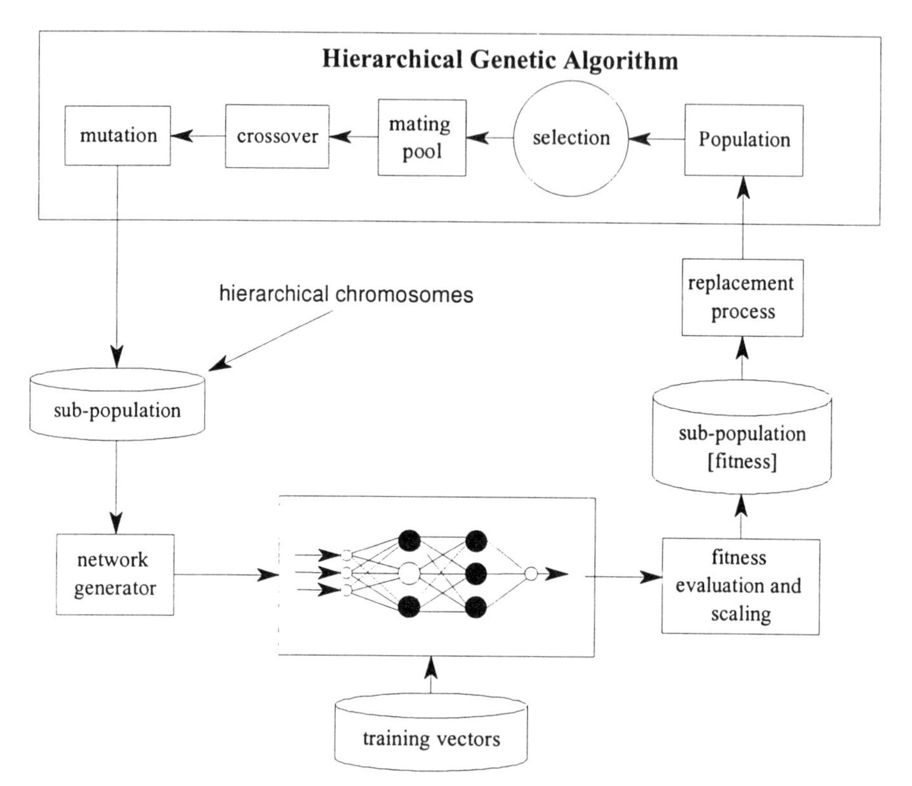

Fig. 7.4. Block Diagram of the Overall HGANN Operation

$$P_{max} = \lambda M \tag{7.4}$$

In the initialization stage, there are $P^{(0)} \leq P_{max}$ chromosomes to be generated. Once a new generation has been produced, new chromosomes are inserted into the population pool to explore the next possible generation of topology.

Objective Functions. The objective of training the network is to minimize two different parameters: the accuracy of the network (f_1) and the complexity of the network (f_2) which is simply defined by the number of active connections in the network. This is calculated based upon the summation of the total number of active connections taking place. The accuracy of the network (f_1) is defined as:

$$f_1 = \frac{1}{N} \sum_{i=1}^{N} (\hat{y}_i - y_i)^2 \tag{7.5}$$

wher N is the size of the testing vector; $\hat{y}_i$ and y_i are the network output and desired output for the i-th pattern of the test vector respectively.

Selection process. Parent Selection is a routine to emulate the survival-of-the-fittest mechanism of nature. Chromosomes in the population are selected for the generation of new chromosomes (offspring) by the certain selection schemes. It is expected that a better chromosome will receive a higher number of offspring and thus has a higher chance of surviving in the subsequent generation. Since there are two different objective functions, (f_1) and (f_2) of the network optimization process, the fitness value of chromosome z is thus determined:

$$f(z) = \alpha \cdot rank[f_1(z)] + \beta \cdot f_2(z) \tag{7.6}$$

where α is accuracy weighting coefficient; β is complexity weighting coefficient; and $rank[f_1(z)] \in Z^+$ is the rank value.

The selection rate of a chromosome z, $tsr(z)$, is determined by:

$$tsr(z) = \frac{F - f(z)}{(size[P^{(k)}] - 1) \cdot F} \tag{7.7}$$

where F is the sum of the fitness value of all chromosomes.

Considering that the accuracy of the network is of paramount importance rather than the complexity of the network, the rule of thumb of the design is such that the weighting coefficients, α, and β take the form as follows:

Let M be the maximum active number of connections in the neural network system, then

$$f_2(z) \leq M, \quad \forall z \in P \tag{7.8}$$

Assuming that at least one successful network has been learnt in the population P, i.e. $\exists z_i \in P$, such that $f_1(z_i) = 0$ and $rank\,[f_1(z_i)] = 1$, then

$$\begin{aligned} f(z_i) &= \alpha + \beta \cdot f_2(z_i) \\ &\leq \alpha + \beta \cdot M \end{aligned} \tag{7.9}$$

where $\beta \in \Re^+$.

Consider that chromosome $z_j \in P$ is failed in learning, i.e. $f_1(z_j) > 0 \Rightarrow rank\,[f_1(z_j)] \geq 2$,

$$\begin{aligned} f(z_j) &= \alpha \cdot rank\,[f_1(z_j)] + \beta \cdot f_2(z_j) \\ &\geq 2\alpha + \beta \cdot f_2(z_j) \\ &> 2\alpha \end{aligned} \tag{7.10}$$

Hence, α is set as following to ensure $f(z_j) > f(z_i)$,

$$\alpha > \beta \cdot M \tag{7.11}$$

Genetic Operations. Since there are two types of genes in the chromosome structure, see Fig. 7.3, specific genetic operations have been designed to suit their purposes. For each type of gene, there are two genetic operations, i.e. crossover and mutation which are recommended.

Control Genes Crossover. A modified multilayer one point crossover operation is applied into the control genes with the probability rate p_{cb}. Once the probability test has passed (a randomly generated number, r_1, is smaller than p_{cb}), one-point crossover is performed in each layer as shown in Fig. 7.5. Parents are separated into two portions by a randomly defined crosspoint at each level. The new control genes are then formed by combining the first part of the parent 1 and the second part of the parent 2 as indicated in Fig. 7.5.

Connection Genes Crossover. Since the connection gene is a vector of real parameters, a one-point crossover operation can thus be directly applied. The operation rate is assumed to be p_{cr}. If a randomly generated number, r_2, is smaller than p_{cr}, the new gene is mated from the first portion of the parent 1 and the last portion in the parent 2 as shown in Fig. 7.6.

Control Genes' Mutation. Bit Mutation is applied for the control genes in the form of a bit-string. This is a random operation that occasionally (with probability p_{mb}, typically 0.01-0.05) occurs which alters the value of a string position so as to introduce variations into the chromosome. Each bit of the control gene is flipped if a probability test is satisfied (a randomly generated number, r_3, is smaller than p_{mb}). An example of Control Genes' Mutation is demonstrated in Fig. 7.7.

Connection Genes mutation. A real value mutation has been designed for the connection genes. For each connection gene, a Gaussian noise is added with probability p_{mr} which can be randomly assigned, (typically 0.05-0.1). The new mutation function is thus:

parent 1
parent 2
High-level Level 1
Level 2
Level n
new chromosome

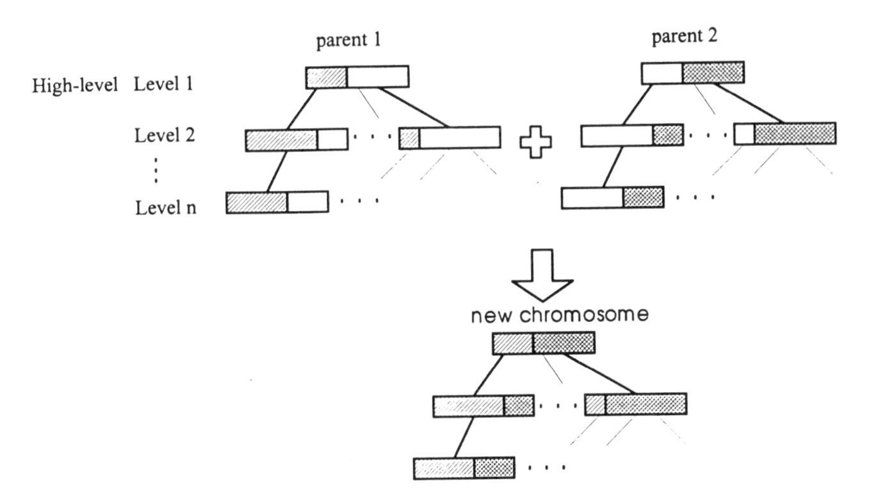

Fig. 7.5. Control Genes' Multilevel One-point Crossover Operation

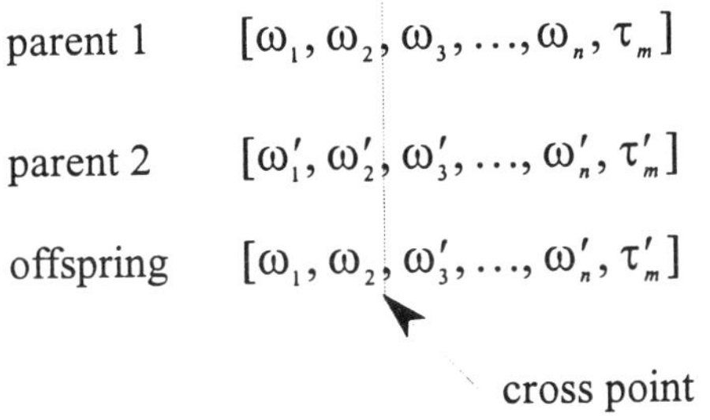

parent 1 $[\omega_1, \omega_2, \omega_3, \ldots, \omega_n, \tau_m]$
parent 2 $[\omega_1', \omega_2', \omega_3', \ldots, \omega_n', \tau_m']$
offspring $[\omega_1, \omega_2, \omega_3', \ldots, \omega_n', \tau_m']$
cross point

Fig. 7.6. Connection Genes' Crossover Operation

0 1 0 0 1 0 1
0 1 1 0 0 0 1

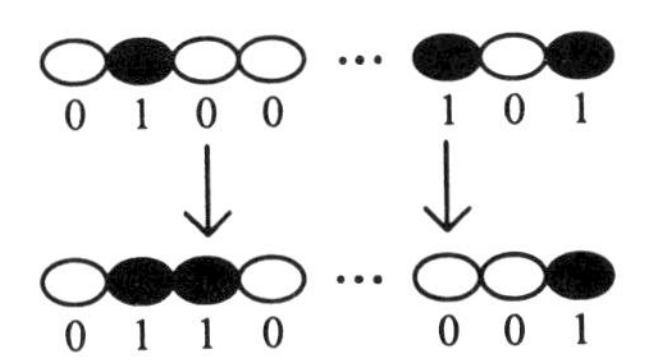

Fig. 7.7. Connection Genes' Mutation Operation

$$m_r(x) = x + N(0,1) \tag{7.12}$$

where x is the current connection weight, and $N(\mu, \sigma^2)$ is a Gaussian random variable with mean μ and variance σ^2.

Insertion Strategy. The top level description of the insertion strategy for the new chromosome z is expressed in Table 7.1:

Table 7.1. Insertion Strategy

Step 1:	If $\{G^{(k)}_{f_2(z)} = \phi$ or $size \left[G^{(k)}_{f_2(z)}\right] < \lambda\}$ then $\{G^{(k+1)}_{f_2(z)} = G^{(k)}_{f_2(z)} \cup \{z\}$ and goto step 3 $\}$ else goto step 2
Step 2:	If $\{f_1(z) < f_m = max\{f_1(z_i), \forall z_i \in G^{(k)}_{f_2(z)}\}\}$ then $\{G^{(k+1)}_{f_2(z)} = \{z_i : f_1(z_i) < f_m, z_i \in G^{(k)}_{f_2(z)}\} \cup \{z\}\}$ else goto step 3
Step 3:	Exit

7.1.3 Simulation Results

To verify the performance of the proposed HGANN system, a subset of suggesting testing functions (a and b) in [122] and an arbitrarily logic function (c) to assess the NN have been used. The following three different 3-input Boolean functions have been introduced for the verification of the HGANN:

(a) Test 1 : XOR
(b) Test 2 : Parity check
(c) Test 3 : Arbitrarily set Logic Function

$$input = \{x_1, x_2, x_3\}; \quad output = x_2 \wedge (x_1 \vee x_3)$$

The genetic operational parameters are shown in Tables 7.2 and 7.3.

Table 7.2. Parameters for Genetic Operations of HGANN

Population Size	20
Generation Gap	1.0
Selection	Roulette Wheel Selection on Rank
Reinsertion	Table 7.1

Table 7.3. Parameters for Chromosome Operations of HGANN

	Control Genes	Connection Genes
Representation	Bit Representation (1 bit)	Real Number
Crossover	One point Crossover	One point Crossover
Crossover Rate	1.0	1.0
Mutation	Bit Mutation	Random Mutation
Mutation Rate	0.05	0.1

For each of above three test cases, 30 trials were conducted to allow comparison of the performance for a single layer GA* (GA) and Back-Propagation (BP) [44] against the HGANN. The basic topology for learning is depicted in Fig. 7.8. It should be noticed that BP was applied to this base topology for all three tests. In the case of the GA and HGANN, both the topology and the connection weights were optimized. The number of chromosome levels and the gene length for HGANN were both set to two. Simulation results are tabulated in Tables 7.4-7.6.

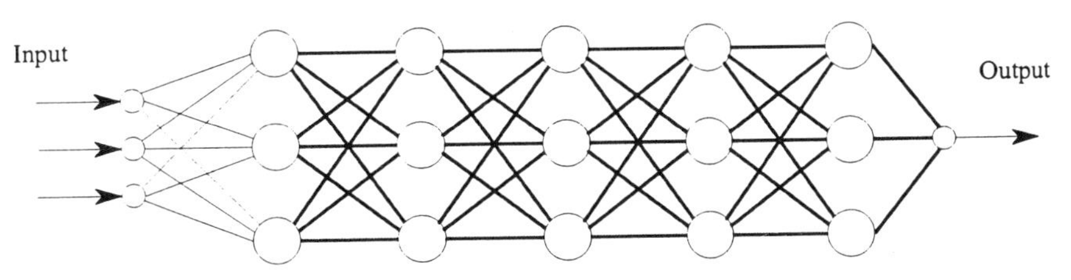

Fig. 7.8. Basic Topology for Learning

Table 7.4. Mean of Terminated Iterations in 30 Trials

Test	HGANN	GA	BP
1	513	1121	1652 [1]†
2	870	2228	1354, [2]
3	37	57	293, [0]

The medium simulation results of the three tests for different algorithms are depicted in Figs. 7.9-7.11. The corresponding topologies obtained after 1000 iterations are shown in Figs. 7.12-7.14. It can be observed that the HGANN has a faster convergence rate than the other two methods for

* Single layer GA has a chromosome structure without any layer control genes.

† Number of trial failed in network training after 10,000 iteration. Terminated iteration number 10,000 is assigned.

Table 7.5. Best of Terminated Iterations in 30 Trials

Test	HGANN	GA	BP
1	38	134	359
2	187	435	279
3	5	6	220

Table 7.6. Standard Deviation of Terminated Iterations in 30 Trials

Test	HGANN	GA	BP
1	431	652	1782
2	870	869	2345
3	28	52	84

all three simulation tests. Moreover, the number of connections was also minimized concurrently in HGANN.

This new scheme has the ability to optimize the NN topology concurrently so that it includes the number of layers and neurons, as well as the associated connection weightings. As a result, the final obtained NN is optimal and is considered to be much more effective for tackling practical problems. Judging from the simulation results, this new computational architecture gives a better performance than techniques using BP and traditional GA methods in terms of the number of training iterations.

7.1.4 Application of HGANN on Classification

Despite the successful demonstrations as shown above, using the benchmark problems that were often used in the 1980s, these results are not convincing enough to justify this approach for practical implementation. A number of deficiencies that have to be overcome for a realistic application exist:

1. all of these problems are purely synthetic and have strong prior regularities in their structure;
2. For some, it is unclear how to measure these in a meaningful way according to the generalization capabilities of a network with respect to the problem.
3. Most of the problems can be solved absolutely, which is untypical for realistic settings.

Therefore, in order to fully justify the capability of the HGANN, a real-life application is introduced. The HGANN technique is used to determine decisions about breast cancer, so that a correct diagnosis can be made for the classification of a tumor as either being benign or malignant.

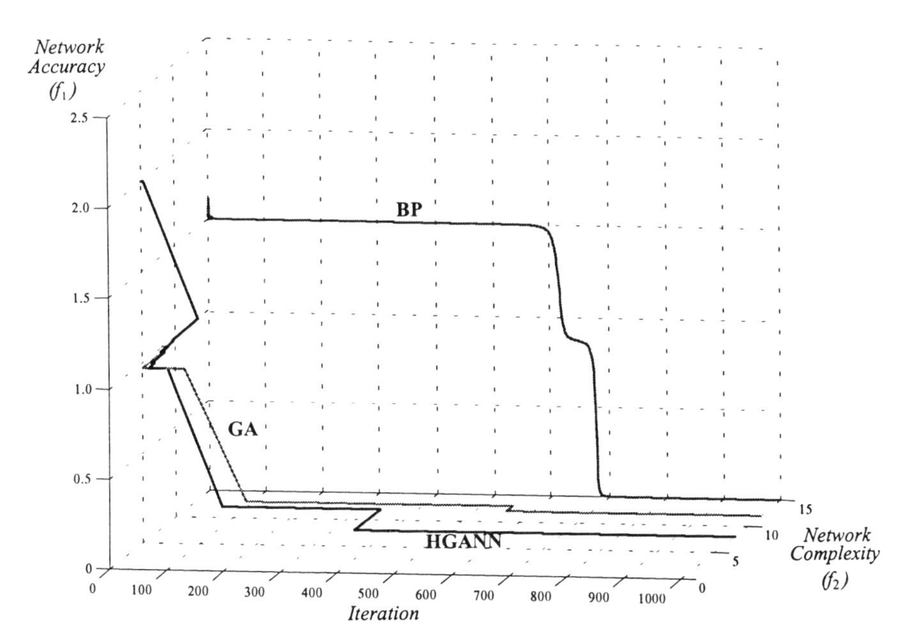

Fig. 7.9. Median Performance in Test 1

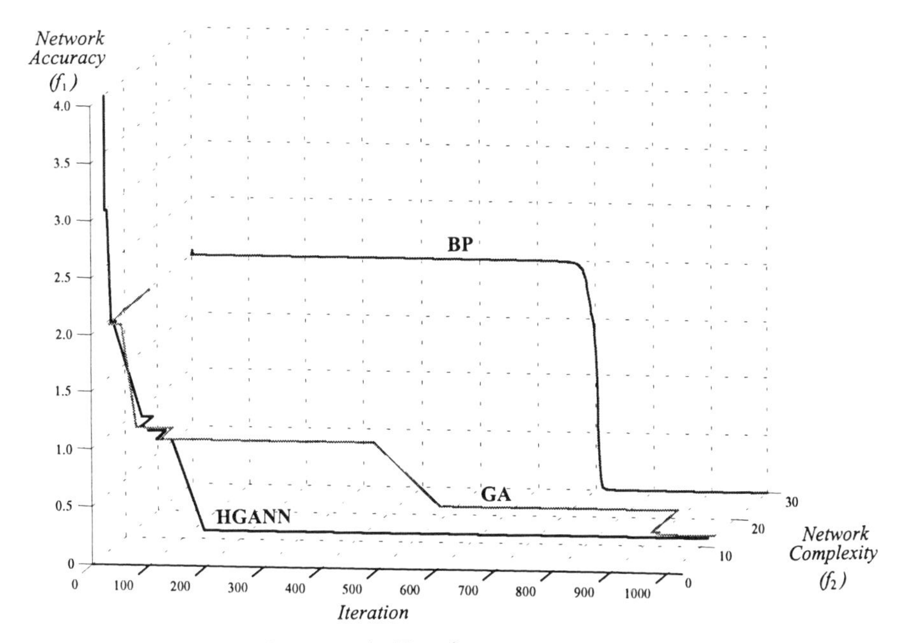

Fig. 7.10. Median Performance in Test 2

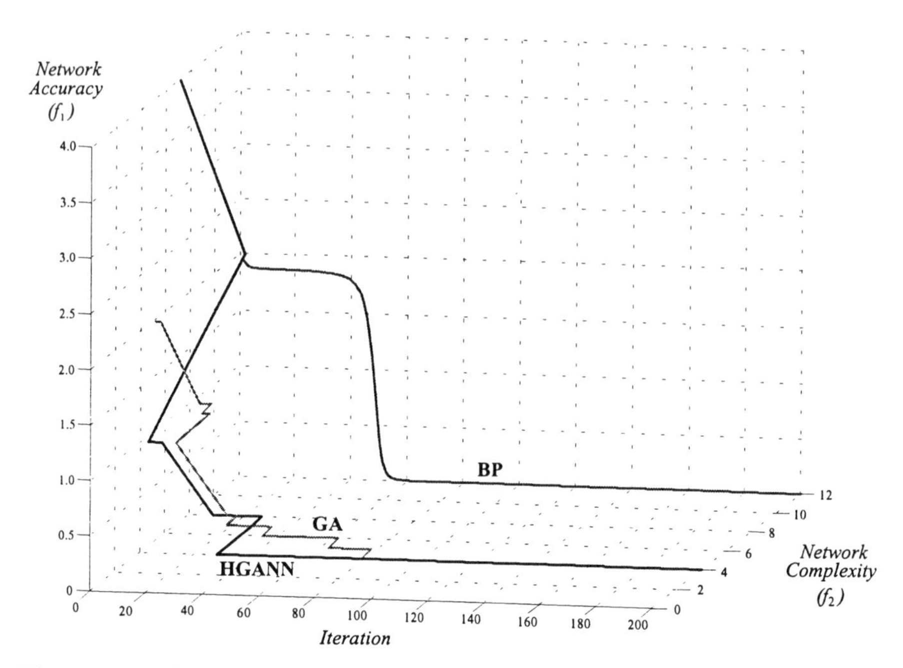

Fig. 7.11. Median Performance in Test 3

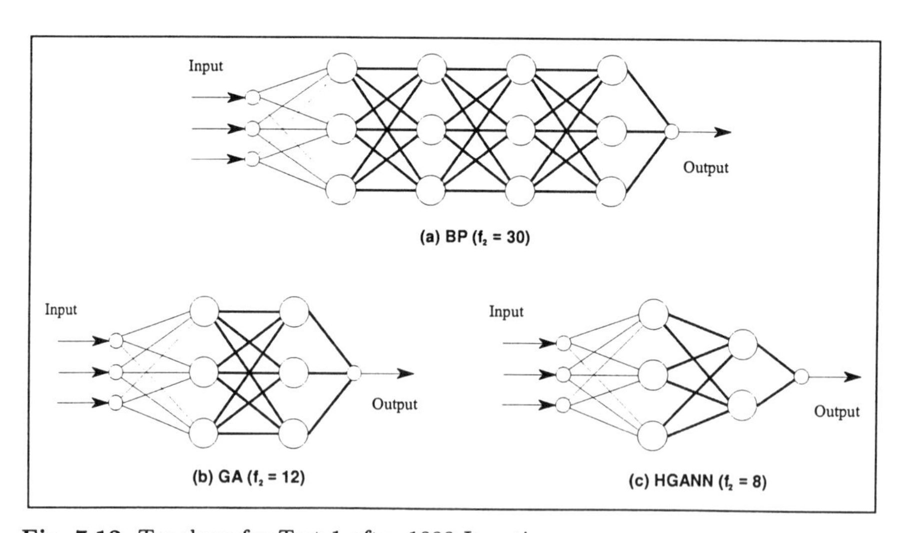

Fig. 7.12. Topology for Test 1 after 1000 Iterations

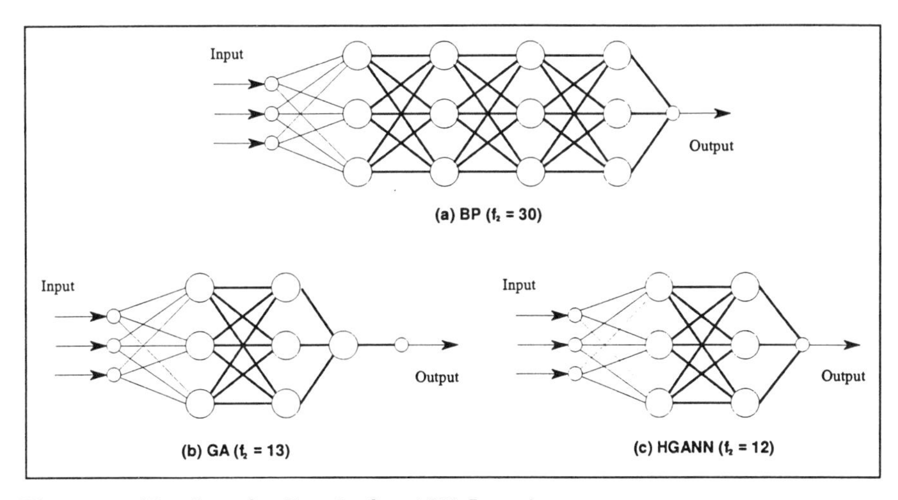

Fig. 7.13. Topology for Test 2 after 1000 Iterations

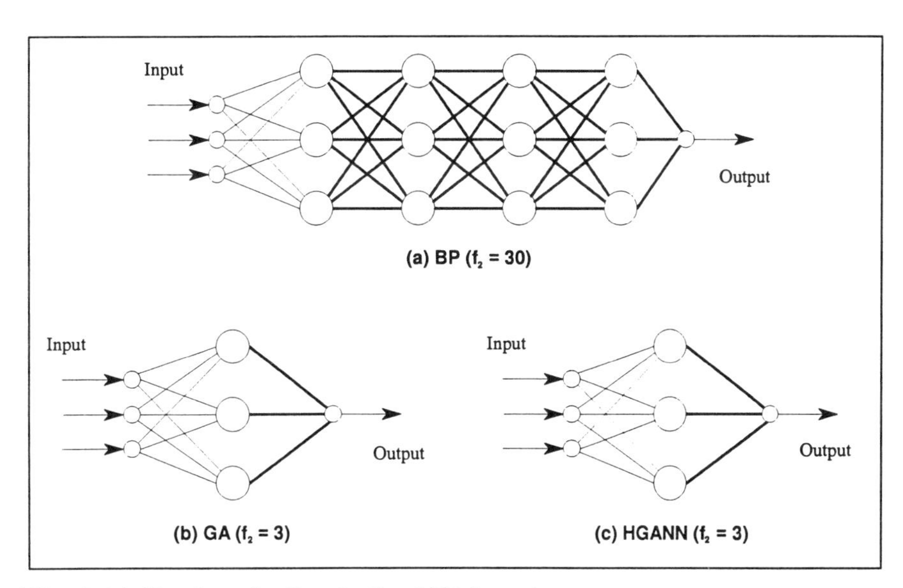

Fig. 7.14. Topology for Test 3 after 1000 Iterations

To facilitate this exercise, the original data was obtained from the University of Wisconsin Hospitals, Madison [95]. There are nine input attributes and two outputs. The input attributes are based on cell descriptions gathered by microscopic examination, of which are listed in Table 7.7.

The data in continuous form were rescaled, first by a linear function so that a mapping into the range of 0...1 could be made. There are 699 examples in total where the class distribution is as follows:

Table 7.7. Input Attributes and Their Domain

Attribute	Domain
Clump Thickness	0-10
Uniformity of Cell Size	0-10
Uniformity of Cell Shape	0-10
Marginal Adhesion	0-10
Single Epithelial Cell Size	0-10
Bare Nuclei	0-10
Bland Chromatin	0-10
Normal Nucleoli	0-10
Mitoses	0-10

Table 7.8. Class Distribution

Class	benign	malign	total
Total number	458	241	699
Total percentage	65.5	34.5	100

In real life, the collected data may comprise some missing attribute values. But they are filled by fixed values which were taken to be the mean of the non-missing values of this attribute. Within this data set, there are 16 missing values for 6-th attribute. So here, these are simply encoded as 0.3, since the average value of the attributes is roughly 3.5.

To use these data for the learning of the neural network learning algorithms, the data must be split into at least two parts:

1. one part for NN training, called the *training data*; and
2. another part for NN performance measurement, called the *test set.*

The idea is that the performance of a network on the test set estimates its performance for real use. It turns out that there is absolutely no information about the test set examples or the test set performance of the network available during the training process; otherwise the benchmark is invalid. The data are, now, classified as

1. 100 training data
2. 599 testing data

The fitness function is defined as:

$$f = \sum_{k=1}^{100} |y_t(k) - r_t(k)| \tag{7.13}$$

where $y_t(k)$ and $r_t(k)$ are the network output and expected output of the training data, respectively.

The error measure (g) is defined as:

$$g = \sum_{k=1}^{599} |y_s(k) - r_s(k)| \tag{7.14}$$

where $y_s(k)$ and $r_s(k)$ are the network output and expected output of the testing data, respectively.

To proceed with the HGANN formulation, we can design the fundamental network as shown in Fig. 7.15. Since there are only two outputs, the corresponding chromosome structure can thus be formed.

This is a NN architecture that has nine-input and one-output. The "1" signifies the output as being benign. By the use of the similar GA operational parameters as before, the performance of HGANN due to the number of identified active neurons is obtained as shown in Figs. 7.16. It can be clearly seen that only four neurons would be necessary instead of 36 neurons (a full connected network) as before. This is a greatly simplified network which is depicted in Fig. 7.17. Based on the error measurement where $g = 26.9$ was obtained, the final HGANN topology will have the accuracy of about 95.5%. This is a slightly better comparable result to that which has already been obtained (93.5% and 93.7%) by [163, 169]. However, in this case, this simple architecture will be much more computationally efficient as compared with the origianl network (Fig. 7.15). As a result, this could potentially lead to a faster procedure for breast cancer diagnosis.

7.2 Fuzzy Logic

Ever since the very first introduction of the fundamental concept of fuzzy reasoning by Zadeh [167] in 1973, its use in engineering disciplines has been widely studied. It has been reported that over 1,000 commercial and industrial

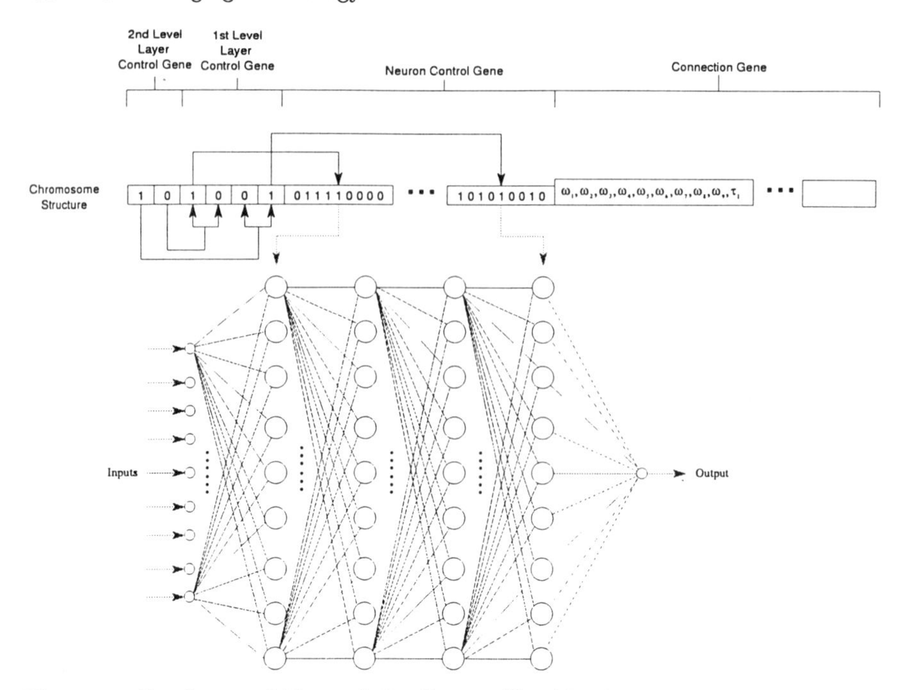

Fig. 7.15. Fundamental Network for Cancer Classification

fuzzy systems have been successfully developed in the space of last few years [106].

Its main attraction undoubtedly lies in the unique characteristics that fuzzy logic systems possess. They are capable of handling complex, nonlinear and sometimes mathematically intangible dynamic systems using simple solutions. Very often, fuzzy systems may provide a better performance than conventional non-fuzzy approaches with less development cost.

However, to obtain an optimal set of fuzzy membership functions and rules is not a easy task. It requires time, experience and skills of the operator for the tedious fuzzy tuning exercise. In principle, there is no general rule or method for the fuzzy logic set-up, although a heuristic and iterative procedure [113] for altering the membership functions to improve performance has been proposed, albeit that this is not optimal. Recently, many researchers have considered a number of intelligent schemes for the task of tuning the fuzzy set. The noticeable neural network approach [76] and the compatible GA methods [71, 79, 80, 110, 142] to optimize the membership functions and rules have become a trend for future fuzzy logic system development. It is our belief that the GA approach to optimize the fuzzy set is sound and that

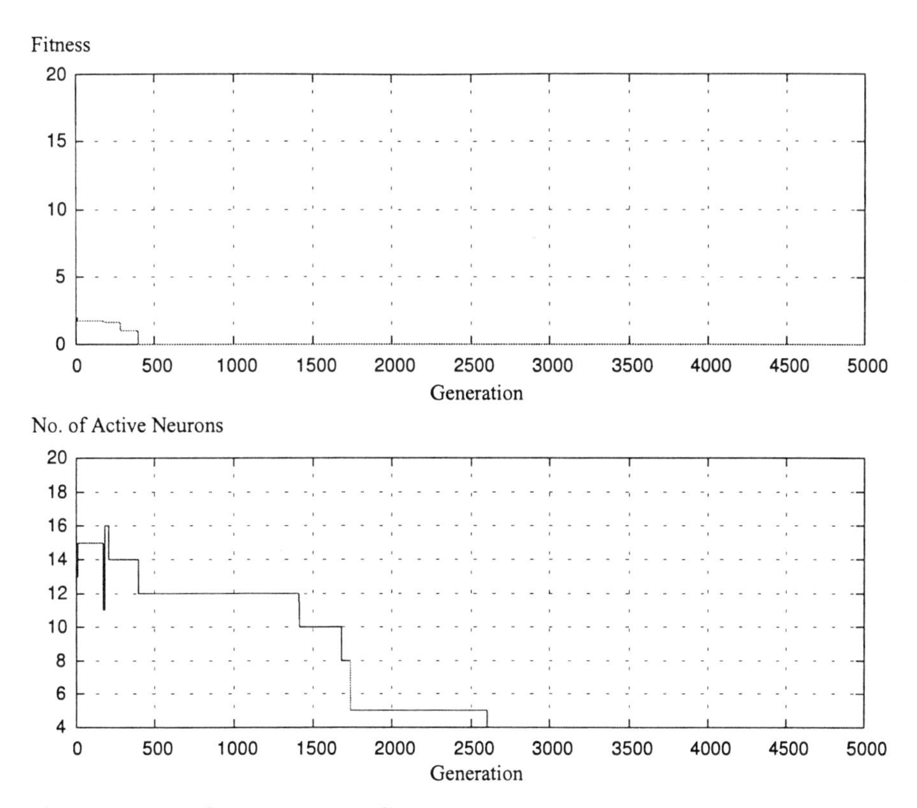

Fig. 7.16. Best Chromosome vs Generation

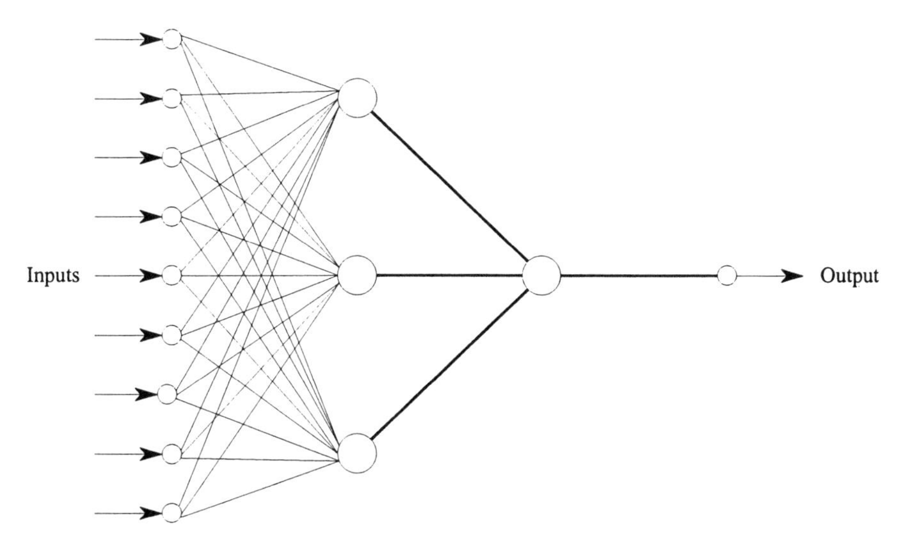

Fig. 7.17. Topology after 5000 Iterations

efforts in this direction should be continued.

Here, another innovative scheme is recommended. This approach differs from the other techniques in that it has the ability to reach an optimal set of memberships and rules without a known overall fuzzy set topology. This can be done only via the attributes of the HGA as discussed before. During the optimization phase, the membership functions need not be fixed. Throughout the genetic operations, a reduced fuzzy set including the number of memberships and rules will be generated. It is the purpose of this section to outline the essence of this technique based on fuzzy control.

7.2.1 Basic Formulation of Fuzzy Logic Controller

The fundamental framework of any fuzzy control system can be realized as shown in Fig. 7.18.

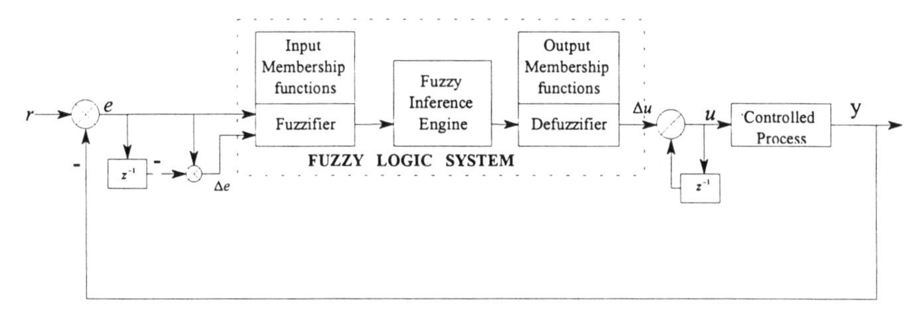

Fig. 7.18. Block Diagram of Genetic Fuzzy Logic Controller

The operational procedure of the fuzzy logic controller (FLC) examines the receiving input variables e and Δe in a fuzzifying manner so that an appropriate actuating signal is derived to drive the system control input (u) in order to meet the ultimate goal of control. The favourable characteristic of the FLC lies is its ability to control the system without knowing exactly how the system dynamics behave. The degree of success using this method relies totally upon the judgment on the error signal e and Δe, where e is defined as $(e = r - y)$ and Δe is the digital rate of the change of e. The basic FLC arrangement is thus depicted in Fig. 7.19.

Judging from the signals $(e, \Delta e)$, a mapping from $\underline{x} = (e, \Delta e) \in X \subset \Re^2$ to $\Delta u \in U \subset \Re$ can be performed. This process is generally known as *Fuzzification.* During the process, each input is classified into fuzzy subsets. Consider the error fuzzy set E as an example, this can be further divided into seven fuzzy subsets $(\mu_i^{(E)})$, defined as *Negative Large (NL,*$(\mu_1^{(E)})$*), Negative Medium (NM,* $(\mu_2^{(E)})$*), Negative Small (NS,*$(\mu_3^{(E)})$ *), Zero (ZE,*$(\mu_4^{(E)})$*),*

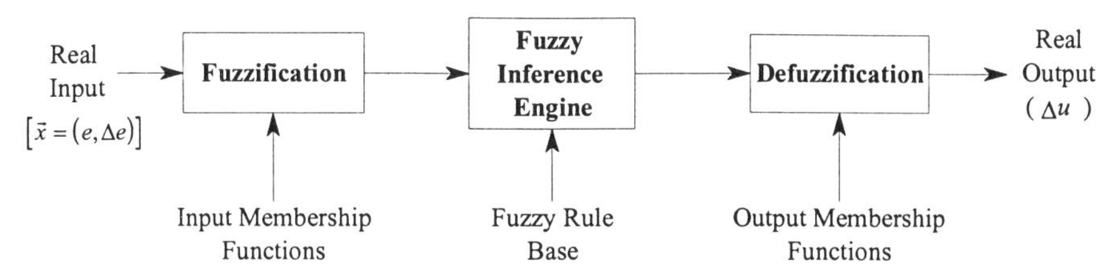

Fig. 7.19. Fuzzy Logic System

Positive Small $(PS,(\mu_5^{(E)}))$, *Positive Medium* $(PM,(\mu_6^{(E)}))$, *Positive Large* $(PM,(\mu_7^{(E)}))$. In general, these subsets can be constructed in the form of triangular membership functions as indicated in Fig. 7.20 in the dynamic range of $[e_{min}, e_{max}]$ as the minimum and maximum magnitude for signal e, respectively.

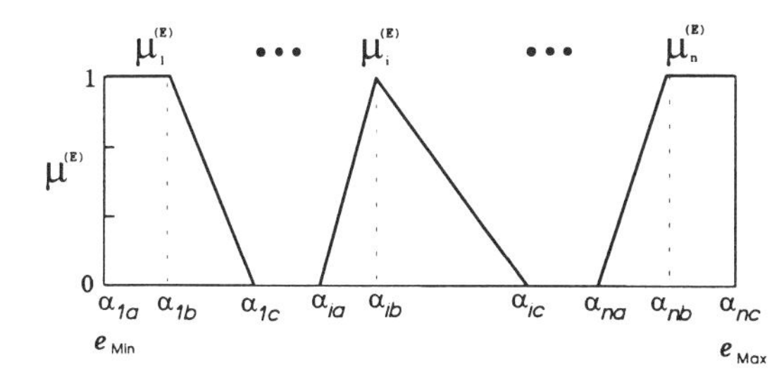

Fig. 7.20. Membership Functions for Fuzzy Set E

The membership functions are defined as follows:

$$\mu_1^{(E)}(e) = \begin{cases} 1 & e \le \alpha_{1b} \\ \frac{\alpha_{1c}-e}{\alpha_{1c}-\alpha_{1b}} & \alpha_{1b} < e < \alpha_{1c} \\ 0 & e \ge \alpha_{1c} \end{cases}$$

$$\mu_i^{(E)}(e) = \begin{cases} \frac{e-\alpha_{ia}}{\alpha_{ib}-\alpha_{ia}} & \alpha_{1a} < e \le \alpha_{1b} \\ \frac{\alpha_{1c}-e}{\alpha_{1c}-\alpha_{1b}} & \alpha_{1b} < e < \alpha_{1c} \\ 0 & e \le \alpha_{ia} \text{ or } e \ge \alpha_{1c} \end{cases} \quad \text{where } i = 2, \ldots, 6$$

$$\mu_n^{(E)}(e) = \begin{cases} 0 & e \le \alpha_{na} \\ \frac{e-\alpha_{na}}{\alpha_{nb}-\alpha_{na}} & \alpha_{nb} < e < \alpha_{na} \\ 0 & e \ge \alpha_{nb} \end{cases} \quad \text{where } n = 7 \tag{7.15}$$

Next, the degree of truth through the input membership functions is obtained and the same method applies to the membership functions for error rate fuzzy set (ΔE) and output fuzzy set (ΔU). Once the membership functions are installed, the fuzzy output (Δu) can be derived. This is commonly obtained by developing a set of fuzzy control rules which are capable of highlighting the concepts of fuzzy implication. The general form of the fuzzy rule is the *"IF $\cdots$ and $\cdots$ THEN $\cdots$"* structure.

Example 7.2.1. If (e is *NS* and Δe is *PS*) *then* Δu is *ZE*.
where *NS* and *PS* are the fuzzy subsets for e and Δe respectively; and *ZE* is the fuzzy subset for Δu.

Once the membership functions have been defined as indicated by Eqn. 7.15, (traditionally, this is normally done by hand and requires a good deal of experience), the appropriate fuzzy rules to govern the system can thus be developed. A typical fuzzy rule table is shown in Fig. 7.21.

e \ Δe	NL	NM	NS	ZE	PS	PM	PL
NL	*NL*	*NL*	*NL*	*NM*	*NS*	*NS*	*ZE*
NM	*NL*	*NM*	*NM*	*NS*	*NS*	*ZE*	*PS*
NS	*NL*	*NM*	*NM*	*NS*	*ZE*	*PS*	*PS*
ZE	*NM*	*NM*	*NS*	*ZE*	*PS*	*PM*	*PM*
PS	*NS*	*NS*	*ZE*	*PS*	*PM*	*PM*	*PL*
PM	*NS*	*ZE*	*PS*	*PM*	*PM*	*PM*	*PL*
PL	*ZE*	*PS*	*PS*	*PM*	*PL*	*PL*	*PL*

Fig. 7.21. IF-THEN Rule

The rules $(R)_{i=1\to 7, j=1\to 7})$ for driving the system input Δu, as shown in Fig. 7.21, are then coded in the following manner:

$R_{1,1}$: If e is *NL* and Δe is *NL* then Δu is *NL*
$R_{1,2}$: If e is *NL* and Δe is *NM* then Δu is *NL*
$R_{1,3}$: If e is *NL* and Δe is *NS* then Δu is *NL*

$\vdots$

$R_{7,7}$: If e is *PL* and Δe is *PL* then Δu is *PL*

or

$$R = R_{1,1} \cup R_{1,2} \cup \cdots \cup R_{7,7} \tag{7.16}$$

Within these 49 individual rules, each rule is capable of governing the fuzzy output Δu. This then allows a wide variation for any of these rules to be valid at the same time. This phenomenon can be illustrated further by the following example.

Example 7.2.2. When a truth value of error signal (e) reaches a degree of truth to 0.7 on *NM* and 0.3 on *NS*; while the error rate signal Δe at the same time touches a degree of truth on 0.5 of *NS* and 0.9 of *ZE*, the associated governing rules can be obtained as indicated by the highlighted rules in Fig. 7.21. These rules are stated as

$R_{2,3}$: If e is *NM* and Δe is *NS* then Δu is *NM*
$R_{2,4}$: If e is *NM* and Δe is *ZE* then Δu is *NS*
$R_{3,3}$: If e is *NS* and Δe is *NS* then Δu is *NM*
$R_{3,4}$: If e is *NS* and Δe is *ZE* then Δu is *NS*

These can be directly mapped into the shaded output membership functions, *NM* and *NS*, for control, as shown in Fig. 7.22.

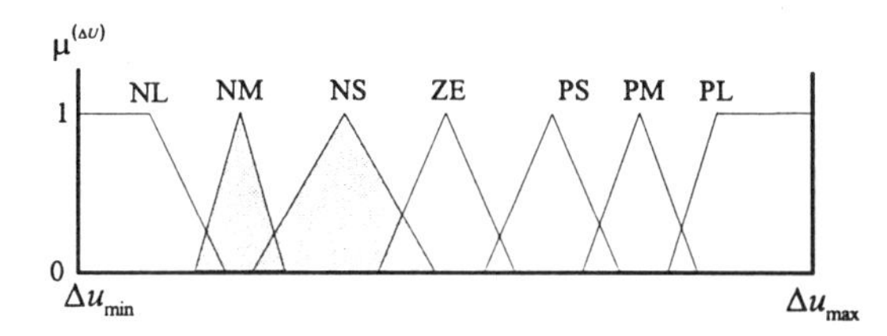

Fig. 7.22. Output Membership Functions

A union operation (Minimum Inferencing),

$$\mu_{(A \cap B)} = \min\left(\mu_A, \mu_B\right) \tag{7.17}$$

where A and B are the fuzzy subsets may apply to determine the degree of truth on the output subsets of $R_{2,3}$, $R_{2,4}$, $R_{3,3}$ and $R_{3,4}$. The Minimum Inferencing on rule $R_{2,3}$ is used as a typical example to illustrate this point as shown in Fig. 7.23. The process should also be repeated for $R_{2,4}$, $R_{3,3}$ and $R_{3,4}$.

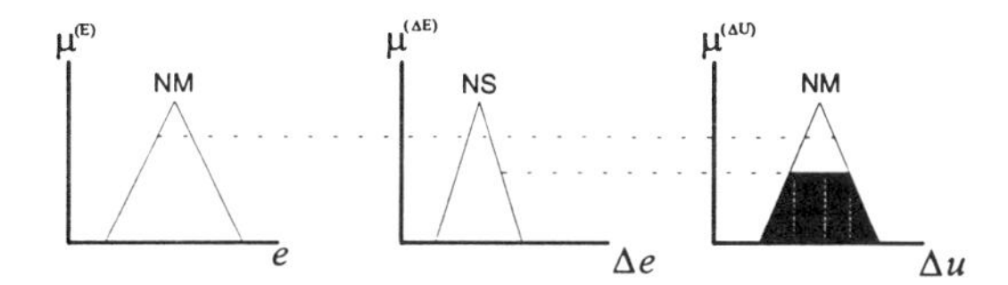

Fig. 7.23. Minimum Inferencing on Rule $R_{2,3}$

In this way, the degree of truth for the output fuzzy subsets can thus be obtained, and these are tabulated in Table 7.9 for the four rules $R_{2,3}$, $R_{2,4}$, $R_{3,3}$ and $R_{3,4}$.

Here, it is clear that more than one degree of truth value can be assigned for an output fuzzy subset. These are, in this case, *NM* (0.3,0.5) or *NS* (0.7,0.3). To ensure a correct decision, a process of interaction (*Maximum composition*) may apply for the combination of NM and NS. This is indicated in Table 7.10.

Finally, it was found that the degrees of truth for the output fuzzy subsets, *NM* and *NS* were 0.5 and 0.7, respectively. From this, the crisp value of the

Table 7.9. Example of Minimum Inferencing

	Error Fuzzy Subset	Error Rate Fuzzy Subset	Minimum Inferencing	Output Fuzzy Subset
$R_{2,3}$	NM (0.7)	NS (0.5)	$\min\{0.7, 0.5\}$	NM (0.5)
$R_{2,4}$	NM (0.7)	ZE (0.9)	$\min\{0.7, 0.9\}$	NS (0.7)
$R_{3,3}$	NS (0.3)	NS (0.5)	$\min\{0.3, 0.5\}$	NM (0.3)
$R_{3,4}$	NS (0.3)	ZE (0.9)	$\min\{0.3, 0.9\}$	NS (0.3)

Table 7.10. Example of Maximum Composition

	$R_{2,3}$	$R_{2,3}$	$R_{2,3}$	$R_{2,3}$	Max Composition	Degree of Truth
NM	0.5	-	0.3	-	$\max\{0.5, 0.3\}$	0.5
NS	-	0.7	-	0.3	$\max\{0.7, 0.3\}$	0.7

output variable Δu_o can be calculated via a process of *defuzzification* as shown in Fig. 7.24.

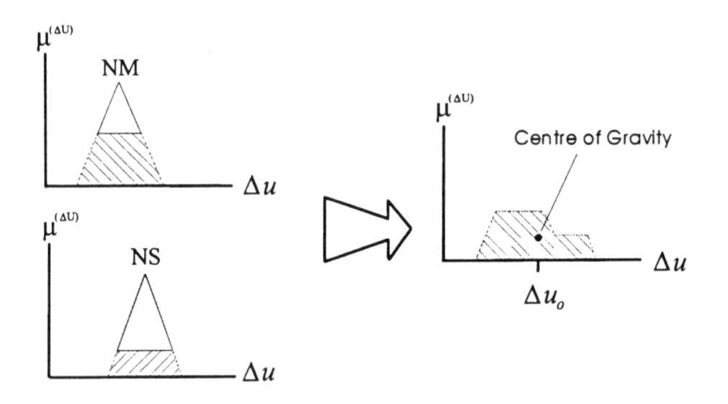

Fig. 7.24. Centre-of-Gravity Defuzzification

The actual value of Δu_o is calculated by the Centroid method:

$$\Delta u_o = \frac{\sum_i \Delta u_i \cdot \mu_i^{(\Delta U)}(\Delta u_i)}{\sum_i \mu_i^{(\Delta U)}(\Delta u_i)} \tag{7.18}$$

where $\mu_i^{(\Delta U)}$ is the membership function of fuzzy subset of ΔU.

This final value of Δu_o will be used to drive the system input (u) for the ultimate system control.

7.2.2 Hierarchical Structure

Having now learnt the complicated procedures of designing FLC, a practical realization of this system is not easy to determine. The dynamic variation of fuzzy input membership functions and the interpretation of governing rules for fuzzy output are the main stumbling blocks to this design. Manually operating procedures for these variables might not only yield a sub-optimal performance, but could also be dangerous if the complete fuzzy sets were wrongly augmented.

Considering that the main attribute of the HGA is its ability to solve the topological structure of an unknown system, then the problem of determining the fuzzy membership functions and rules could also fall into this category. This approach has a number of advantages:

- an optimal and the least number of membership functions and rules are obtained;
- no pre-fixed fuzzy structure is necessary; and
- simpler implementing procedures and less cost are involved.

Hence, it is the purpose of this section to introduce the HGA for the designing of FLC [147]. The conceptual idea is to have an automatic and intelligent scheme to tune the fuzzy membership functions and rules, in which the closed loop fuzzy control strategy remains unchanged, as indicated in Fig. 7.25.

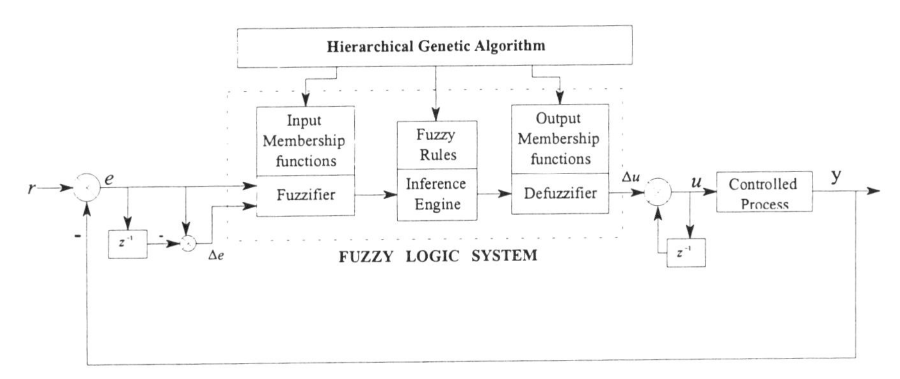

Fig. 7.25. HGA Fuzzy Logic Control System

Chromosome of HGA. Similar to other uses of the HGA, the hierarchical chromosome for the FLC structure must be correctly formulated. In this case, the chromosome of a particular fuzzy set is shown in Fig. 7.26. The chromosome consists of the usual two types of genes, the control genes and parameter genes. The control genes, in the form of bits, determine the membership function activation, whereas the parameter genes (similar to those stated in Eqn. 7.15) are in the form of real numbers to represent the membership functions.

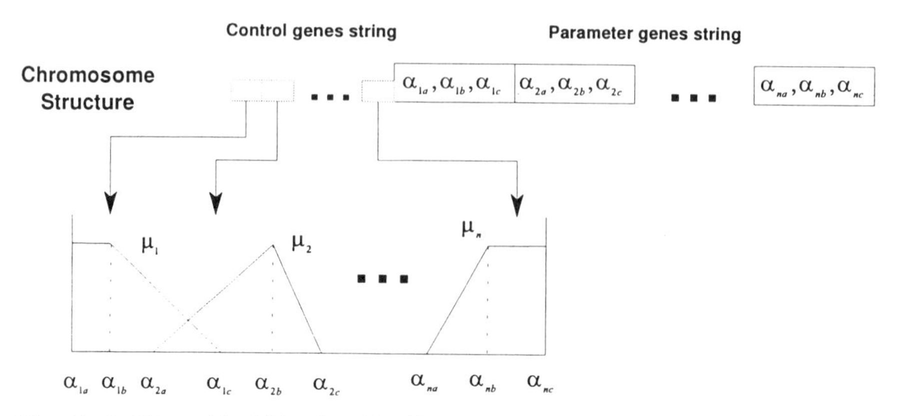

Fig. 7.26. Hierarchical Membership Chromosome Structure

With the two input fuzzy sets of error signals, (e), error rate (Δe) and the output fuzzy set of Δu, we can thus construct the overall membership chromosome structure as in Fig. 7.27.

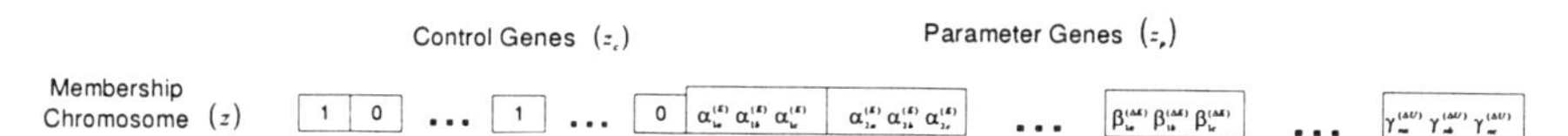

Fig. 7.27. HGA Chromosome Structure

The parameter genes (z_p) of the membership chromosome take the form:

$$z_p = \Big\{ \alpha_{1a}^{(E)}, \alpha_{1b}^{(E)}, \alpha_{1c}^{(E)}, \ldots, \alpha_{ma}^{(E)}, \alpha_{mb}^{(E)}, \alpha_{mc}^{(E)}, \beta_{1a}^{(\Delta E)}, \beta_{1b}^{(\Delta E)}, \beta_{1c}^{(\Delta E)}, \ldots, \beta_{na}^{(\Delta E)}, \beta_{nb}^{(\Delta E)}, \beta_{nc}^{(\Delta E)}, \gamma_{1a}^{(\Delta U)}, \gamma_{1b}^{(\Delta U)}, \gamma_{1c}^{(\Delta U)} \ldots \gamma_{pa}^{(\Delta U)}, \gamma_{pb}^{(\Delta U)}, \gamma_{pc}^{(\Delta U)} \Big\}$$

where m, n and p are the maximum allowable number of fuzzy subset of E, ΔE and ΔU, respectively; $\alpha_{ia}^{(E)}, \alpha_{ib}^{(E)}, \alpha_{ic}^{(E)}$ define the input membership function of i-th fuzzy subset of E; $\beta_{ja}^{(\Delta E)}, \beta_{jb}^{(\Delta E)}, \beta_{jc}^{(\Delta E)}$ define the input membership function of j-th fuzzy subset of ΔE; and $\gamma_{ka}^{(\Delta U)}, \gamma_{kb}^{(\Delta U)}, \gamma_{kc}^{(\Delta U)}$ define the output membership function of k-th fuzzy subset of ΔU.

To obtain a complete design for the fuzzy control design, an appropriate set of fuzzy rules is required to ensure system performance. At this point, it should be stressed that the introduction of the control genes is done to govern the number of fuzzy subsets E, ΔE and ΔU. As a result, it becomes impossible to set a universal rule table similar to the classic FLC approach for each individual chromosome. The reason for this is that the chromosomes may vary from one form to another, and that each chromosome also has a different number of fuzzy subsets.

Therefore, the fuzzy rules based on the chromosome set-up should be classified. For a particular chromosome, there should be w, x and y active subsets with respect to E, ΔE and ΔU in an independent manner. This can be represented by a rule table, as shown in Table 7.11, with a dimension $w \times x$. Then, each cell defines a rule for the controller, i.e. the $i - j$ element implies rule $R_{i,j}$:

$R_{i,j}$: If e is E_i and Δe is D_j then Δu is U_k

where E_i, D_j, U_k are now the linguistic name, similar to "Large", "Small" and so on, to characterize the fuzzy subsets of error, error rate and output set, respectively.

Table 7.11. The Rule Base in Tabular Form

	D_1	D_2	$\cdots$	D_j	$\cdots$	D_x
E_1	U_1	U_2		$\cdots$		U_j
E_2	U_2	U_3		$\cdots$		U_j
$\vdots$			$\vdots$			
E_i		$\cdots$		U_k	$\cdots$	
$\vdots$			$\vdots$			
E_w	U_i			$\cdots$		U_y

The Fuzzy Rule Chromosome, $H_{(w,x,y)}$, is then formulated in the form of an integer matrix where

$$H_{(w,x,y)} = \{h_{i,j} : h_{i,j} \in [1, y] \quad \forall i \leq w,\ j \leq x\} \tag{7.19}$$

Example 7.2.3. For a Fuzzy Rule Chromosome with $w = 2$, $x = 2$ and $y = 2$, then,

$$H_{(2,2,2)} = \begin{bmatrix} 1 & 2 \\ 2 & 2 \end{bmatrix} \tag{7.20}$$

From $H_{(2,2,2)}$, there are four rules:

$R_{1,1}$: If e is E_1 and Δe is D_1 then Δu is U_1
$R_{1,2}$: If e is E_1 and Δe is D_2 then Δu is U_2
$R_{2,1}$: If e is E_2 and Δe is D_1 then Δu is U_2

$R_{2,2}$: If e is E_2 and Δe is D_2 then Δu is U_2

Genetic Cycle. Once the formulation of the chromosome has been set for the fuzzy membership functions and rules, genetic operation cycle can be performed. This cycle of operation for the FLC optimization using an HGA is illustrated in Fig. 7.28.

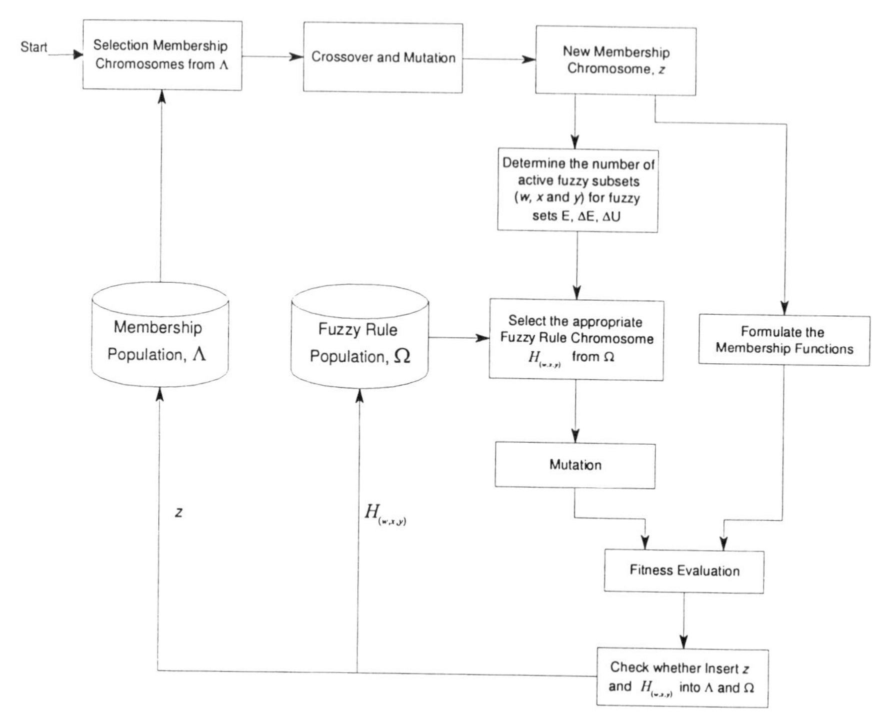

Fig. 7.28. Genetic Cycle for Fuzzy Logic System Optimization

Population. There are two population pools, (Λ) and (Ω), for storing the membership and fuzzy rule chromosomes, respectively. The HGA chromosomes are grouped in Λ, while the fuzzy rule chromosomes are stored in the fuzzy rule population, Ω. Within Ω, there should be a total number of $(m-1) \times (n-1) \times (p-1)$ fuzzy rule sets. However, only one particular single rule set can be matched with $H_{(w,x,y)}$ in order to satisfy the chromosome that possesses the w, x and y active fuzzy subsets of E, ΔE and ΔU, respectively.

Genetic Operations. Considering that there are various types of gene structure, a number of different genetic operations have been designed. For the crossover operation, a one point crossover is applied separately for both the control and parameter genes of the membership chromosomes within certain operation rates. There is no crossover operation for fuzzy rule chromosomes since only one suitable rule set $H_{(w,x,y)}$ can be assisted.

Bit mutation is applied for the control genes of the membership chromosome. Each bit of the control gene is flipped ("1" or "0") if a probability test is satisfied (a randomly generated number, r_c, is smaller than a pre-defined rate). As for the parameter genes, which are real-number represented, random mutation is applied. A special mutation operation has been designed to find the optimal fuzzy rule set. This is a delta shift operation which alters each element in the fuzzy rule chromosome as follows:

$$h_{i,j} = h_{i+\Delta i,j+\Delta j} \tag{7.21}$$

where Δi, Δj have equal chance to be 1 or -1 with a probability of 0.01.

Fitness Evaluation. Before evaluating the fitness value of the chromosome pair (z, H), their phenotype must be obtained. In some cases, a direct decoding of the membership chromosome may result in invalid membership functions. For example, Fig. 7.29 represents an invalid membership function for error fuzzy set because the ranges $(\alpha_{1c}, \alpha_{3a})$ and $(\alpha_{4c}, \alpha_{7c})$ are unclassified (only the error set is shown for clarity).

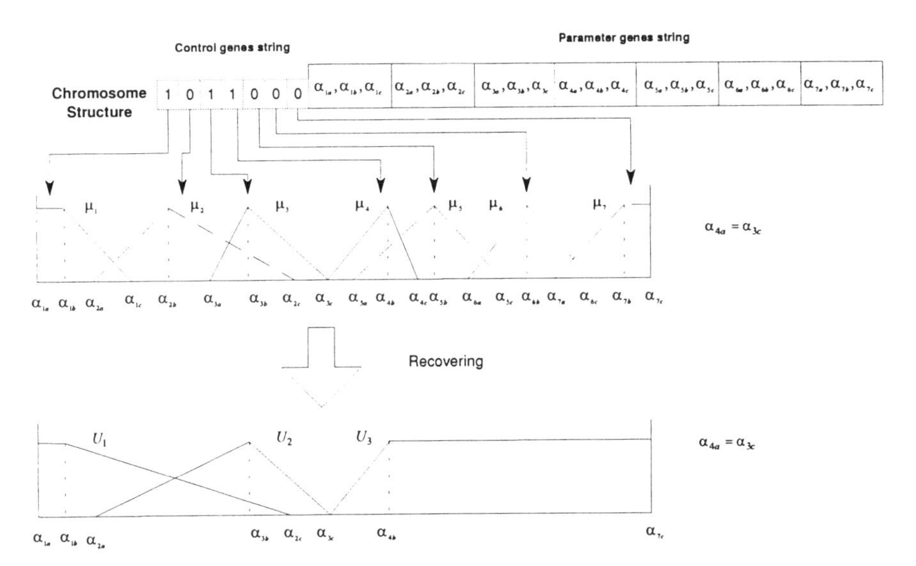

Fig. 7.29. Recovery of Invalid Fuzzy Membership Functions

To ensure that there was no undefined region, a remedial procedure was operated to ensure validation. The decoded fuzzy membership functions were recovered as shown by the final membership characteristics in Fig. 7.29. It should be noted that the parameter gene remained unaltered but merely changed the interpretation of its form. In this way, the complexity of tuning the fuzzy memberships and rules can thus be optimized and the overall structure can be greatly reduced.

Together with the fuzzy rule table generated from the fuzzy rule chromosome, a full set of FLC can then be designed. A fitness value, $f(z, H)$, can then be determined which reflects the fitness of the FLC.

Insertion Strategy. The grouping method, as described in Chap. 5, was adopted in order to find optimal membership functions of E, ΔE and ΔU and the appropriate fuzzy rules. The population of membership chromosomes Λ is divided into several subgroups, $S_{(i,j,k)}$, such that

$$\Lambda = S_{(2,2,2)} \cup S_{(2,2,3)} \ldots \cup S_{(2,n,p)} \cup S_{(3,2,2)} \ldots \cup S_{(m,n,p)} \tag{7.22}$$

and

$$S_{(i,j,k)} \cap S_{(w,x,y)} \neq \emptyset \qquad \forall (i \neq w \wedge j \neq x \wedge k \neq y) \tag{7.23}$$

where $S_{(i,j,k)}$ is the subgroup of chromosome that represents those with i, j and k active fuzzy subsets for E, ΔE and ΔU, respectively.

The maximum number of subgroups in Λ is thus $(m-1)\times(n-1)\times(p-1)$. A concurrent factor, λ (typically assigned as 3-5), is used to define the maximum elements stored in the membership subgroup,

$$size\left[S_{(i,j,k)}\right] \leq \lambda \qquad \forall\, 2 \leq i \leq m,\ 2 \leq j \leq n, 2 \leq k \leq p \tag{7.24}$$

where $size\left[S_{(i,j,k)}\right]$ is the number of elements in $S_{(i,j,k)}$.

Table 7.12 explains the insertion strategy for new membership chromosome (z) with active w, x, and y fuzzy subsets for sets E, ΔE and ΔU, respectively, with new fuzzy rule chromosome, $H_{(w,x,y)}$.

The complete genetic cycle continues until some termination criteria, for example, meeting the design specification or number of generation reaching a predefined value, are fulfilled.

7.2.3 Experimental Results

To test the design of the HGA fuzzy logic controller, an experimental piece of equipment which consisted of a water pump having a 1.5 horse power engine and a water tank was used for the investigation. It simulated the constant pressure booster pump system designed for a water supply system [67].

The compensated actuator unit was the water pump with a variable frequency converter (VFC) attached. The actuating signal came from a pressure sensor placed in a pipe downstream and its output signal was fed back into the VFC to change the pump speed. A schematic diagram of the water supply system is shown in Fig. 7.30.

Table 7.12. Insertion Strategy

At generation $(k+1)$

Step 1:
If $\left\{S_{(w,x,y)} = \emptyset \quad \text{or} \quad size\left[S_{(w,x,y)}\right] < \lambda\right\}$
then
$S_{(w,x,y)} = S_{(w,x,y)} \cup \{z\}$ and
$\Omega = \left\{H_{(i,j,k)} : H_{(i,j,k)} \in \Omega \quad i \neq w, j \neq x, k \neq y\right\} \cup \left\{H_{(w,x,y)}\right\}$
else
goto step 2

Step 2:
If $\left\{f(z, H_{(w,x,y)}) < f_{max} = \max\left\{f(z_i, H_{(w,x,y)}),\ \forall z_i \in S_{(w,x,y)}\right\}\right\}$
then
$S_{(w,x,y)} = \left\{z_i : F(z_i) < f_{max},\ z_i \in S_{(w,x,y)}\right\} \cup \{z\}$ and
$\Omega = \left\{H_{(i,j,k)} : H_{(i,j,k)} \in \Omega \quad i \neq w, j \neq x, k \neq y\right\} \cup \left\{H_{(w,x,y)}\right\}$
else
goto step 3

Step 3:
Exit

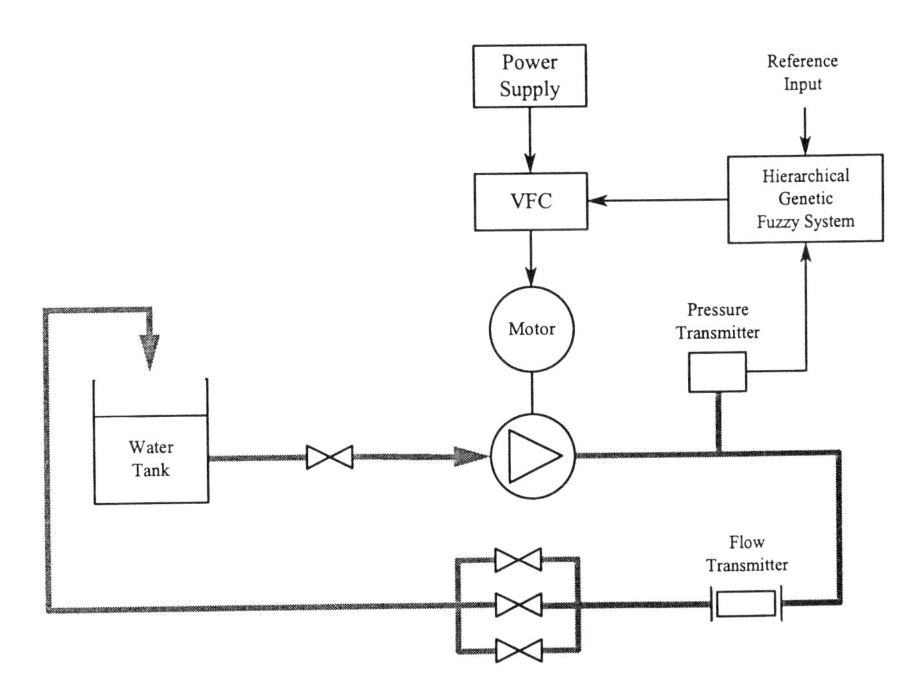

Fig. 7.30. Experimental Apparatus

Table 7.13. Parameters of HGA for Fuzzy Controller

	Membership Chromosome Control Genes	Connection Genes	**Fuzzy Rule Chromosome**
Representation	Binary	Real Number	Integer
Population Size No. of Offspring	20 2		216 (m=n=p=7) –
Crossover Crossover Rate	One-Point 1.0	One-Point 1.0	– –
Mutation Mutation Rate	Bit Mutation 0.02	Random Mutation 0.02	Eqn. 7.21 Eqn. 7.21
Selection	Roulette Wheel Selection on Rank		Based on the no. of active fuzzy subsets
Reinsertion	Table 7.12		Direct replacement

Fuzzy Learning and Optimization. The parameter setting of HGA to optimize the required fuzzy subsets is tabulated in Table 7.13 with a command reference signal in the following form:

$$r(k) = \begin{cases} 1000 & 1 \leq k \leq 400 \\ 1200 & 401 \leq k \leq 800 \\ 1000 & 801 \leq k \leq 1200 \end{cases} \tag{7.25}$$

where k is the sample.

The design specification is to meet two objective functions:

1. Minimum output steady state error

$$f_1 = \frac{1}{100}\left[\sum_{k=701}^{800} (y(k) - r(k))^2 + \sum_{k=1101}^{1200} (y(k) - r(k))^2\right] \tag{7.26}$$

where $r(k)$ and $y(k)$ are the reference and plant outputs, respectively; and

2. Minimum overshoot and undershoot

$$f_2 = p_1 + p_2 \tag{7.27}$$

where

$$\text{overshoot:} \quad p_1 = \begin{cases} \frac{y_{max} - r(401)}{r(401) - r(400)} & y_{max} > r(401) \\ 0 & y_{max} \leq r(401) \end{cases}$$

and

$$\text{undershoot:} \quad p_2 = \begin{cases} \frac{r(801)-y_{min}}{r(800)-r(801)} & y_{min} < r(801) \\ 0 & y_{min} \geq r(801) \end{cases}$$

with $y_{max} = max\{y(k) : 401 \leq k \leq 800\}$ and $y_{min} = min\{y(k) : 801 \leq k \leq 1200\}$

Based on the design procedures that have already been described in the above sections, the obtained plant output response is obtained after 20 generations of computation. The result is shown in Fig. 7.31.

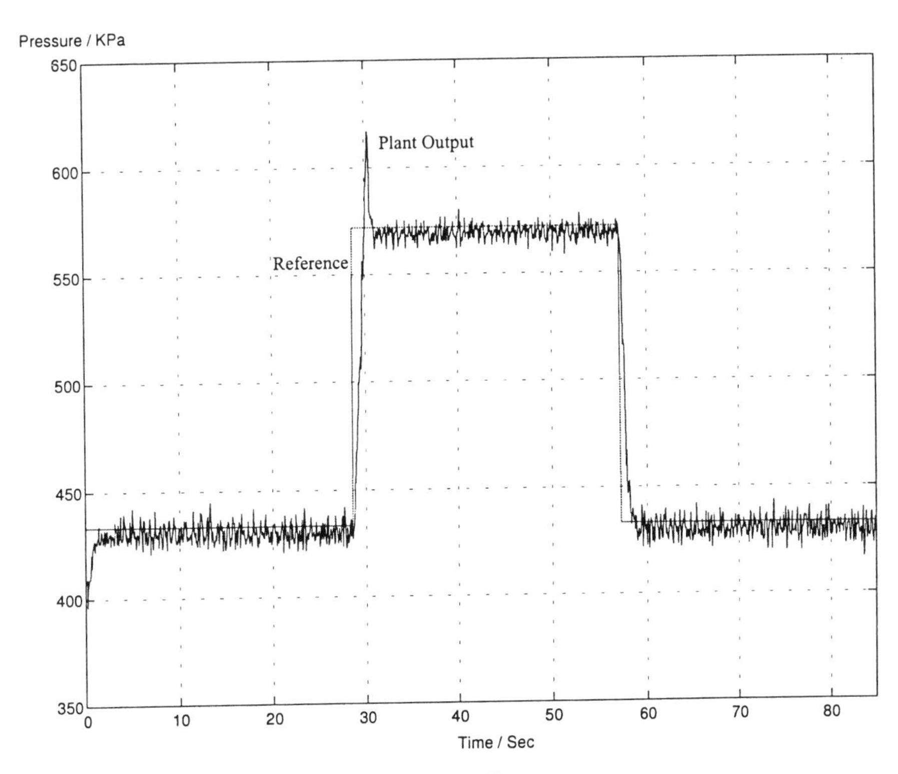

Fig. 7.31. Output Response of the final FLC

The same set of procedures is repeated again, in this case, the maximum allowable overshoot and undershoot are now set to 0%. This result is depicted in Fig. 7.32. These clearly indicate that the HGA fuzzy control design scheme is fully justified. The corresponding FLC design parameters as well as the obtained fuzzy subsets and membership functions (Figs. 7.33, 7.34 and 7.35) are tabulated as follows:

- The membership chromosome is tabulated in Table 7.14 with the associated figures of each fuzzy set.

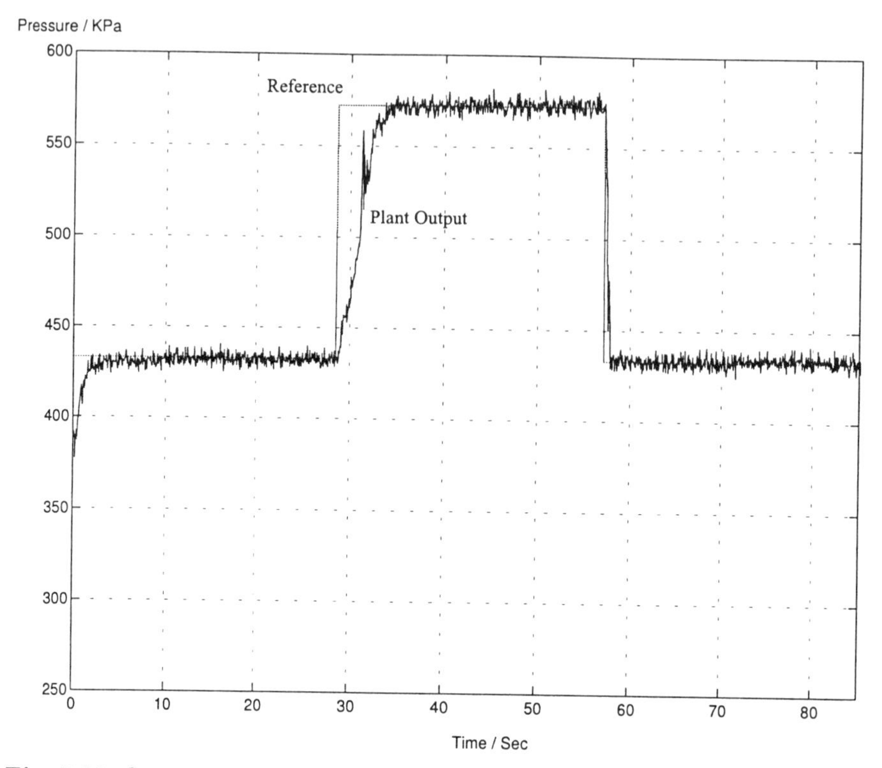

Fig. 7.32. Output Response of the best FLC with Multiobjective

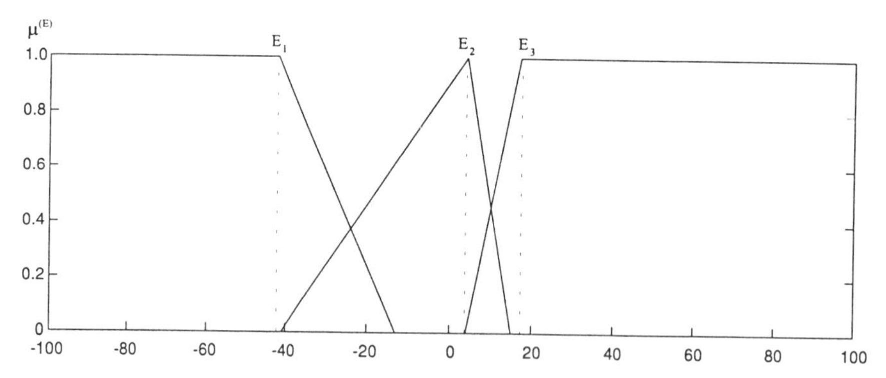

Fig. 7.33. Fuzzy Subsets and Membership Functions of e

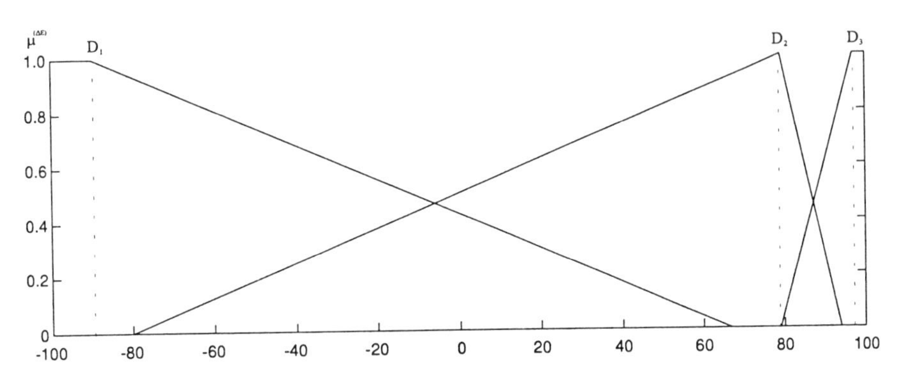

Fig. 7.34. Fuzzy Subsets and Membership Functions of Δe

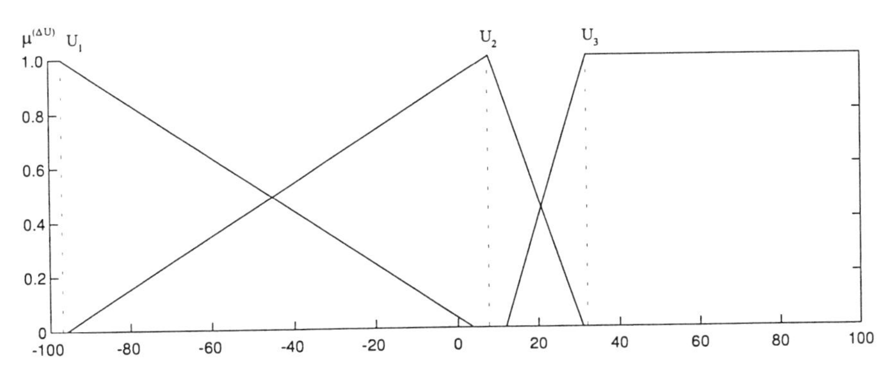

Fig. 7.35. Fuzzy Subsets and Membership Functions of Δu

Table 7.14. Optimal Membership Functions Obtained

Control Genes	Parameter Genes	Fuzzy Subset
$z_c^{(E)} = [0010110]$	$z_p^{(E)} = [(-100, -98, -48), (-97, -47, -45),$ $(-46, -42, -29), (-41, -22, -13),$ $(-19, 4, 15), (4, 17, 40), (29, 47, 100)]$	Fig. 7.33
$z_c^{(\Delta E)} = [1000101]$	$z_p^{(\Delta E)} = [(-100, -90, -66), (-80, -59, -46),$ $(-51, -18, -38), (19, 50, 67),$ $(59, 79, 79), (79, 80, 94), (80, 97, 100)]$	Fig. 7.34
$z_c^{(\Delta U)} = [1001010]$	$z_p^{(\Delta U)} = [(-100, -97, -95), (-96, -75, -65),$ $(-66, -64, 4), (-52, 8, 25),$ $(12, 31, 31), (31, 32, 59), (56, 92, 100)]$	Fig. 7.35

– The fuzzy rule chromosome is obtained as

$$H = \begin{bmatrix} 0 & 1 & 1 \\ 1 & 1 & 2 \\ 1 & 2 & 2 \end{bmatrix}$$

which implies a rule table as in Table 7.15.

Table 7.15. Optimal Rule Table

		Error Rate Fuzzy Set		
		D_1	D_2	D_3
Error	E_1	U_1	U_2	U_2
Fuzzy	E_2	U_2	U_2	U_3
Set	E_3	U_2	U_3	U_3

It has been shown that a reduced size of subsets of fuzzy membership functions and rules is obtained by the use of HGA while still meeting the requirements of system performance. This result is considered to be compatible to those obtained using the conventional fuzzy logic design schemes.

Closed Loop Performance. To further verify the performance of the obtained FLC design, two sets of experimental tests were conducted. Firstly, an irregular square waveform command signal was applied to the water supply system at a number of operating points. It can be seen from Fig. 7.36 that the pressure rise and fall characteristics closely track the set points.

Secondly, a nominal operating point of the water pipe pressure was chosen for dynamic disturbance testing. The disturbance was created by turning the water valve along the pipe to the "on" and "off" state while the water supply was in operation. Fig. 7.37 shows that the pressure recovered well from the disturbance despite the condition of the water valve.

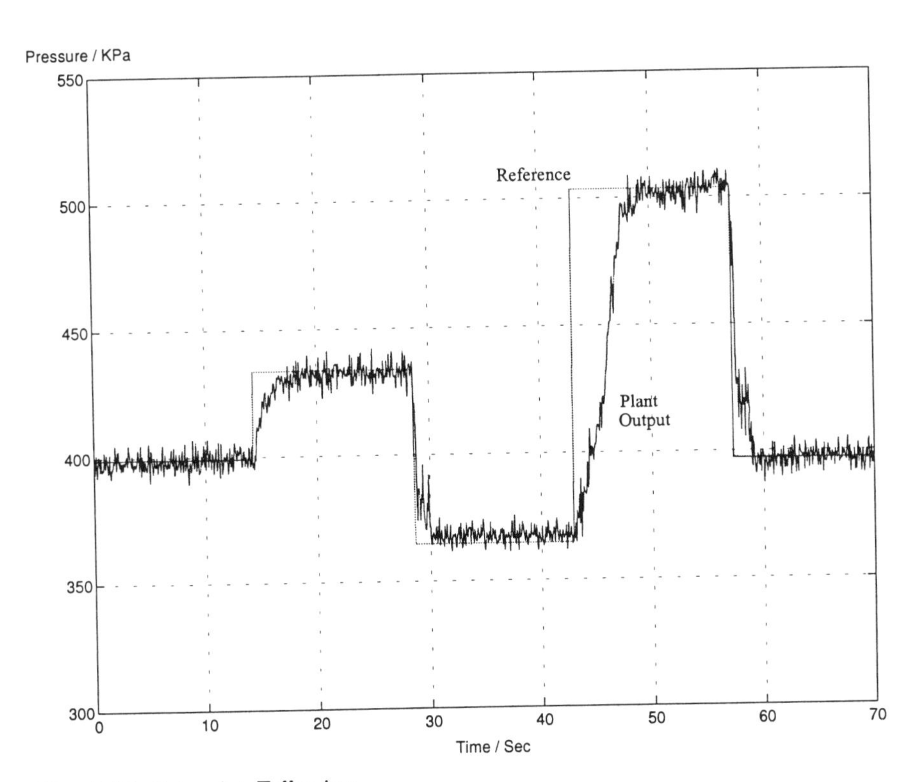

Fig. 7.36. Set-point Following

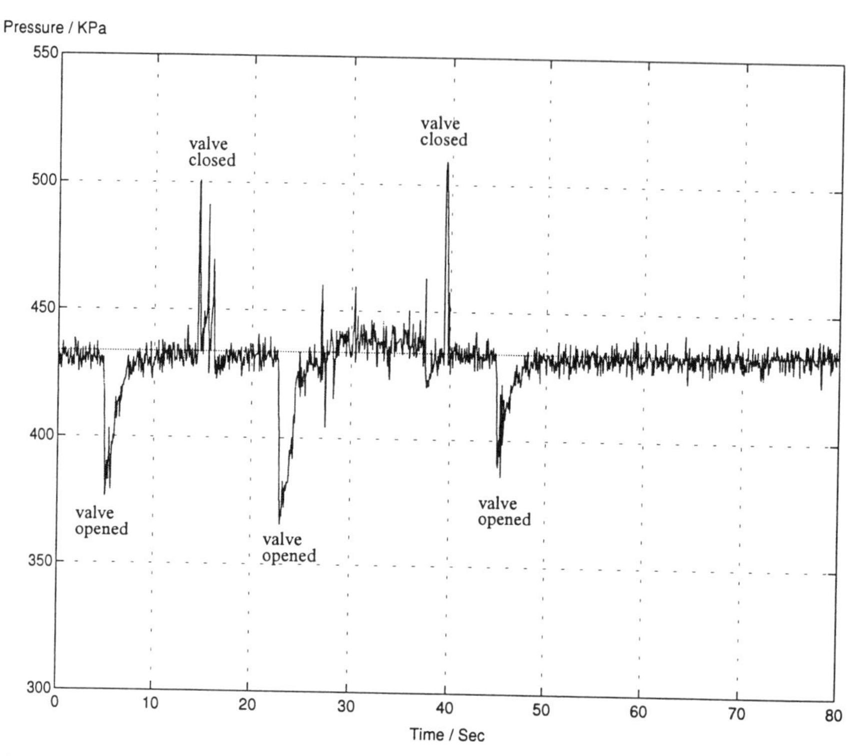

Fig. 7.37. Closed Loop Disturbance

APPENDIX A

Least Mean Square Time Delay Estimation (LMSTDE)

The time delay is estimated by a Finite Impulse Response (FIR) model. The estimation error $e(k)$ is then equal to

$$\begin{aligned} e(k) &= y(k) - AX(k) \\ &= y(k) - \sum_{i=-L}^{L} a_i x(k-i) \end{aligned} \tag{A.1}$$

where $A = \begin{bmatrix} a_{-L} & a_{-L+1} & \cdots & a_L \end{bmatrix}$ is the $(2L+1)$ filter parameter vector; $X(k) = \begin{bmatrix} x(k+L) & x(k+L-1) & \cdots & x(k-L) \end{bmatrix}^T$ is the input vector; and $y(k)$ is the delayed signal.

The filter weights a_i are updated by minimizing the mean square error (MSE) as below:

$$A(k+1) = A(k) + 2\mu_w e(k) X^T(k) \tag{A.2}$$

where μ_w is the gain constant that regulates the speed and stability of adaptation.

APPENDIX B

Constrained LMS Algorithm

The constrained LMS algorithm is formulated as follows:

$$\begin{aligned}\hat{D}(k+1) &= \hat{D}(k) - \mu_n \frac{\partial e^2(k)}{\partial \hat{D}} \\ &= \hat{D}(k) - \mu_n e(k) \sum_{n=-L}^{L} x(k-n) f(n - \hat{D}(k)) \end{aligned} \tag{B.1}$$

where

$$\begin{aligned} e(k) &= y(k) - \sum_{n=-L}^{L} sinc(n - \hat{D}(k)) x(k-n) \\ f(v) &= \frac{cos(\pi v) - sinc(\pi v)}{v} \end{aligned}$$

and μ_n is a convergence factor controlling the stability.

The initial value of $\hat{D}(0)$ must be within the range of $D \pm 1$ so as to retain a unimodal error surface of $e(k)$.

APPENDIX C

Linear Distributed Random Generator

According to the random mutation expressed in Eqn. 2.6, a Guassian distributed random number is added on the genes for mutation. Such a random process is not easy to generate in hardware. However, the design of a pseudo random number generator is simple but possesses a uniform distribution that is not suitable for this application. Therefore, a new method to generate the approximated Guassian distributed random numbers has been proposed by simply manipulating the pseudo random number formulation.

Considering that the output of the random function Ψ in Eqn. 2.6 is formulated as

$$\Psi = \begin{cases} b & \text{if} \quad a > b \\ 0 & \text{else} \end{cases} \tag{C.1}$$

where a, b in this case, are the outputs of two independent pseudo random generators, with the output of each pseudo random generator being set to $(\mu - 3\sigma, \mu + 3\sigma)$. The distribution of the pseudo random number generated is indicated in Fig. C.1b.

In this way, an approximated Gaussian distribution was obtained as shown in Fig. C.1(c). This distribution was found to be realistic and very similar to that obtained from a true Gaussian noise generator as shown in Fig. C.1a.

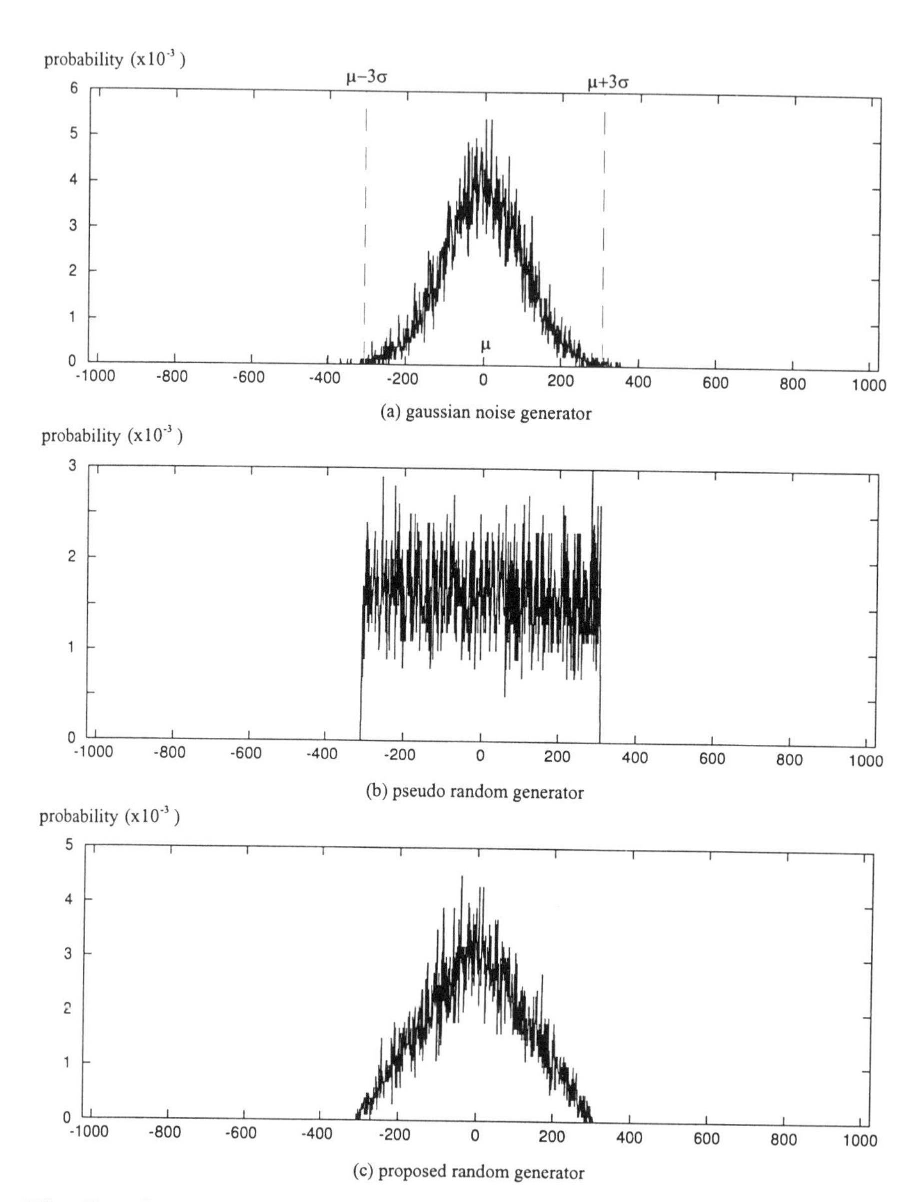

Fig. C.1. Comparison of Random Number Generators

APPENDIX D

Multiplication Algorithm

A high-speed VLSI multiplication algorithm using redundant binary representation was implemented, and signed digit number representation [5] was adopted. This representation was a fixed radix 2 and a digit set $\{\bar{1}, 0, 1\}$ where $\bar{1}$ denotes -1. An n-digits redundant binary integer $Y = [y_{n-1} \cdots y_0]_{SD2}(y_i \in \{\bar{1}, 0, 1\})$ has the value $\sum_{i=0}^{n-1} y_i \times 2^i$.

The multiplier algorithm based on the redundant binary representation [140] is formed by a binary tree of redundant binary adders. Multicand and multiplier are converted into equivalent redundant binary integers and then, an amount of n n-digit partial products represented in redundant binary representation is generated. The computations can be performed in a constant time independent of n.

The partial products were added up in pairs by means of a binary tree of redundant binary adders and the product represented in the redundant binary representation was obtained. The addition of two numbers in the redundant binary number system can be carried out in a constant time independent of the word length of operands. The constant time addition of two redundant binary numbers can be realized by the Carry-Propagation-Free Addition (CPFA). The CPFA is performed in two steps:

1. it is used to determine the intermediate carry $c_i(\in \{\bar{1}, 0, 1\})$ and the intermediate sum digit $s_i(\in \{\bar{1}, 0, 1\})$ at each position, which satisfies the equation $x_i + y_i = 2c_i + s_i$, where x_i and y_i are the augend and addend digits. There are six types of combinations of the two digits in addition as tabulated in Table D.1; and
2. the sum digit $z_i(\in \{\bar{1}, 0, 1\})$ at each position is obtained by adding the intermediate sum digit s_i and the intermediate carry c_{i-1} from the next-lower-order position, without generating a carry at any position in the second step. As a result, the additions are performed in a time proportional to $log_2 n$.

Table D.1. Computation Rules For CPFA

Type	Combination $\{x_i, y_i\}$	next-lower-order-position $\{x_{i-1}, y_{i-1}\}$	carry c_i	sum s_i
1	$\{1, 1\}$	———	1	0
2	$\{1, 0\}$	Both are nonnegative	1	$\bar{1}$
		Otherwise	0	1
3	$\{1, \bar{1}\}$	———	0	0
4	$\{0, 0\}$	———	0	0
5	$\{0, \bar{1}\}$	Both are nonnegative	0	$\bar{1}$
		Otherwise	$\bar{1}$	1
6	$\{\bar{1}, 1\}$	———	$\bar{1}$	0

Finally, the product must be converted into binary representation. As an n-digit redundant binary number

$$A \left(= \sum_{i=0}^{n-1} a_i \times 2^i, a_i \in \{\bar{1}, 0, 1\} \right)$$

is equal to

$$A^+ \left(= \sum_{a_i=1} a_i \times 2^i \right) - A^- \left(= \sum_{a_i=\bar{1}} (-a_i) \times 2^i \right)$$

Therefore, a conversion of an n-digit redundant binary integer into the equivalent $(n+1)$-bit 2's complement binary integer is performed by subtracting A^- from A^+, where A^- and A^+ are n bit unsigned binary integers from the positive digits and the negative digits in A, respectively. This conversion can be performed in a time proportional to $log_2 n$ by means of an ordinary carry-look-ahead adder. In addition, the extended Booth's algorithm can be applied to further reduce the computation time and the amount of hardware required.

APPENDIX E

Digital IIR Filter Designs

Based on the HGA formulation as indicated in Chap. 6, the design of digital filters in the form of LP, HP, BP and BS are thus possible. The genetic operational parameters are shown in Tables E.1 and E.2

Table E.1. Parameters for Genetic Operations

Population Size	100
Generation Gap	0.2
Selection	Multiobjective Ranking
Reinsertion	Replace the lowest Rank
sflag = 1	$f_1 = 0 \wedge f_2 = 0$
N_{max}	20000

Table E.2. Parameters for Chromosome Operations

	Control Genes	Coefficient Genes
Representation	Bit Representation (1 bit)	Real Number Representation
Crossover	Normal Crossover	Normal Crossover
Crossover Rate	0.85	0.8
Mutation	Bit Mutation	Random Mutation
Mutation Rate	0.15	0.1

The fundamental structure of $H(z)$ which applies to all four filters is given as:

$$H(z) = K \prod_{i=1}^{3} \frac{(z + b_i)}{(z + a_i)} \prod_{j=1}^{4} \frac{(z^2 + b_{j1}z + b_{j2})}{(z^2 + a_{j1}z + a_{j2})} \tag{E.1}$$

The control genes (g_c) and coefficient genes (g_r) in this case are thus

$$g_c \in B^{14} \tag{E.2}$$

$$g_r = \left\{ \begin{array}{l} a_1, a_2, a_3, a_{11}, a_{12}, a_{21}, a_{22}, a_{31}, a_{32}, a_{41}, a_{42}, \\ b_1, b_2, b_3, b_{11}, b_{12}, b_{21}, b_{22}, b_{31}, b_{32}, b_{41}, b_{42} \end{array} \right\} \tag{E.3}$$

where $B = [0, 1]$ and the ranges of $a_i, b_i, a_{i1}, a_{i2}, b_{i1}, b_{i2}$ are defined as in Table 6.2.

The design criteria for the filters is tabulated in Table E.3.

Table E.3. Summary of Filter Performances

Filter Type	Design Criteria		K
LP	$0.89125 \le \lvert H(e^{j\omega})\rvert \le 1,$	$0 \le \lvert\omega\rvert \le 0.2\pi$	$\lvert H(1)\rvert = 1$
	$\lvert H(e^{j\omega})\rvert \le 0.17783,$	$0.3\pi \le \lvert\omega\rvert \le \pi$	
HP	$\lvert H(e^{j\omega})\rvert \le 0.17783,$	$0 \le \lvert\omega\rvert \le 0.7\pi$	$\lvert H(e^{j\pi})\rvert = 1$
	$0.89125 \le \lvert H(e^{j\omega})\rvert \le 1,$	$0.8\pi \le \lvert\omega\rvert \le \pi$	
BP	$\lvert H(e^{j\omega})\rvert \le 0.17783,$	$0 \le \lvert\omega\rvert \le 0.25\pi$	$\lvert H(e^{0.5\pi j})\rvert = 1$
		$0.75\pi \le \lvert\omega\rvert \le \pi$	
	$0.89125 \le \lvert H(e^{j\omega})\rvert \le 1,$	$0.4\pi \le \lvert\omega\rvert \le 0.6\pi$	
BS	$\lvert H(e^{j\omega})\rvert \le 0.17783,$	$0.4\pi \le \lvert\omega\rvert \le 0.6\pi$	$\lvert H(1)\rvert = 1$
	$0.89125 \le \lvert H(e^{j\omega})\rvert \le 1,$	$0 \le \lvert\omega\rvert \le 0.25\pi$	
		$0.75\pi \le \lvert\omega\rvert \le \pi$	

APPENDIX F

Development Tools

The study and evaluation of GA, are essentially non-analytic, largely depending on simulation. While they are strongly application independent, GA software has potentially a very broad domain of application. Part of the common software package is briefly introduced and more information can be found in [65].

Genetic Algorithm Toolbox in MATLAB

A GA Toolbox is developed [18] for MATLAB [98]. Given the versatility of MATLAB's high-level language, problems can be coded in m-files easily. Coupling this with MALTAB's advanced data analysis, visual tools and special purpose application domain toolboxes, the user is presented with a uniform environment with which to explore the potential of GA.

GENESIS

GENEtic Search Implementation System (GENESIS) was developed by John Grefenstette [59]. It is a function optimization system based on genetic search techniques. As the first widely available GA programme, GENESIS has been very influential in stimulating the use of GA, and several other GA packages have been generated because of its capability.

A real number representation and binary representation are allowable. A number of new options have been added, including: a display mode that includes an interactive user interface, the option to maximize or minimize the objective function, the choice of rank-based or proportional selection algorithm, and an option to use a Gray code as a transparent lower level representation.

GENOCOP

GEnetic Algorithm for Numerical Optimization for COnstrained Problems (GENOCOP) was developed by Zbigniew Michalewicz and details can be obtained in [101]. The Genocop system has been designed to find a global optimum (minimum or maximum) of a function with additional linear equalities and inequalities constraints. It runs on any UNIX and DOS system.

GENEsYs

GENEsYs [151] is a GENESIS-based GA implementation which includes extensions and new features for experimental purposes. Different Selection schemes like linear ranking, Boltzmann, (μ, λ)-selection, and general extinctive selection variants are included. Crossover operators and self-adaptation of mutation rates are also possible. There are additional data-monitoring facilities such as recording average, variance and skew of object variables and mutation rates, and creating bitmap-dumps of the population.

TOLKIEN

TOLKIEN (TOoLKIt for gENetics-based applications) ver 1.1 [149] is a C++ class library named in memory of J.R.R. Tolkien. A collection of reusable objects have been developed for genetics-based applications. For portability, no compiler specific or class library specific features are used. The current version has been compiled successfully using Borland C++ Ver. 3.1 and GNU C++. TOLKIEN contains a number of useful extensions to the generic GA. For example:

- chromosomes of user-definable types; binary, character, integer and floating point chromosomes are provided;
- gray code encoding and decoding;
- multi-point and uniform crossover;
- diploidy;
- various selection schemes such as tournament selection and linear ranking
- linear fitness scaling and sigma truncation.

Distributed GENESIS 1.0

Distributed GENESIS 1.0 (DGENESIS) was developed by Erick Cantú-Paz. It is an implementation of migration genetic algorithms, described in Sect. 3.1.2. Its code is based on GENESIS 5.0. Each subpopulation is handled by a UNIX process and communication them is handled with Berkeley sockets.

The user can set the migration rate, migration interval and the topology of the communication between subpopulations in order to realize migration GA.

This version of DGENESIS requires the socket interface provided with 4.2BSD UNIX. It has run successfully on DECStations (running Ultrix 4.2), Sun workstations (with SunOS), microVAXes (running Ultrix 4.1) and PCs (with 386BSD 0.1).

In any network, there are fast and also slow machines. To make the most of available resources, the work load in the participating systems can be balanced by assign to each machine a different number of processes according to their capabilities.

Generic Evolutionary Toolbox

Generic Evolutionary Toolbox (GenET) is a generic evolutionary algorithm, a toolbox for fast development of GA applications and for research in evaluating different evolutionary models, operators, etc.

The package, in addition to allowing for fast implementation of applications and being a natural tool for comparing different models and strategies, is intended to become a depository of representations and operators. Currently, only floating point representation is implemented in the library with few operators.

The algorithm provides a wide selection of models and choices. For example, POPULATION models range from generational GA, through steady-state, to (n,m)-EP and (n,n+m)-EP models (for arbitrary problems, not just parameter OPTIMIZATION). (Some are not finished at the moment). Choices include automatic adaptation of operator probabilities and a dynamic ranking mechanism, etc.

REFERENCES

1. Actel Corporation (1994): FPGA Data Book and Design Guide.
2. Anderson, E.J. and M.C. Ferris (1990): A genetic Algorithm for the assembly line balancing problem. Technical Report TR 926, Computer Sciences Department, University of Wisconsin-Madison.
3. Angeline, P.J., G.M. Saunders, and J.B. Pollack (1994): An evolutionary algorithm that constructs recurrent neural networks. IEEE Trans. Neural Networks, **5(1)**, 54–65.
4. Asakawa, K and Hideyuki Takagi (1994): Neural networks in Japan. Communication of the ACM, **37(3)**, 106–112.
5. A. Avizienis (1961): Signed-digit number representations for fast parallel arithmetic., IEEE Trans. Electron. Comput. **EC-10**, 389–400.
6. Baker, J.E. (1985): Adaptive selection methods for genetic algorithms. Proc. 1st Int Conf on Genetic Algorithms, 101–111.
7. Baker J.E. (1987): Reducing bias and inefficiency in the selection algorithms. Proc. 2nd Int. Conf. Genetic Algorithms. Lawrence Erlbaum Associates, Hillsdale, NJ, 14–21.
8. Baluja, S. (1993): Structure and performance of fine-grain parallelism in genetic search. Proc. 5th Int. Conf. Genetic Algorithm.
9. Beasley, D., D.R. Bull, R.R. Martin (1993): An overview of genetic algorithms: Part 1, fundamentals. University Computing, **15(2)**, 58–69.
10. Beasley, D., D.R. Bull, R.R. Martin (1993): An overview of genetic algorithms: Part 2, research topics. University Computing, **15(4)**, 170–181.
11. Berge, O., K.O. Petterson and S. Sorzdal (1988): Active cancellation of transformer noise: Field measurements. Applied Acoustics, **23**, 309–320.
12. Booker, L. (1987): Improving search in genetic algorithms. Genetic Algorithms and Stimulated Annealing, L. Davis (Eds), 61–73.
13. Braun, H. (1990): On solving travelling salesman problems by genetic algorithms. Proc. First Workshop Parallel Problem Solving from Nature, Springer Verlag, Berlin, 129–133.
14. Cantú-Paz, E. (1995): A summary of research on parallel genetic algorithms. IlliGAL Report No. 95007, Illinois Genetic Algorithms Laboratory, University of Illinois at Urbana-Champaign.
15. Chan, Y.T., J.M. Riley, and J.B. Plant (1981): Modeling of time delay and its application to estimation of nonstationary delays. IEEE Trans. Acoust., Speech, Signal Processing, **ASSP-29**, 577–581.
16. Chen, D., C. Giles, G. Sun, H. Chen, Y. Less, and M. Goudreau (1993): Constructive learning of recurrent neural network. Proc. IEEE Int. Conf. Neural Network **3**, 1196–1201.
17. Cheuk, K.P., K.F. Man, Y.C. Ho and K.S. Tang (1994): Active noise control for power transformer. Proc. Inter-Noise 94, 1365–1368.

18. Chipperfield, A.J., P.J. Fleming and H. Pohlheim (1994): A genetic algorithm toolbox for MATLAB. Proc. Int. Conf. on Systems Engineering, Coventry, UK, 6–8.
19. Chipperfield, A.J. and P.J. Fleming (1994): Parallel genetic algorithms: A survey. ACSE Research Report No. 518, University of Sheffield.
20. Cobb, H.G. (1990): An investigation into the use of hypermutation as an adaptive operator in genetic algorithms having continuous, time-dependent nonstationary environments. NRL Memorandum Report 6760.
21. Cobb, H.G. and J.J. Grefenstette (1993): Genetic algorithms for tracking changing environments. Proc. 5th Int. Conf. Genetic Algorithms, 523–530.
22. Cohoon, J.P., W.N. Martin and D.S. Richards (1991): A multi-population genetic algorithm for solving the k-partition problem on hyper-cubes. Proc. 4th Int. Conf. Genetic Algorithms, 244–248.
23. Daniels R.W. (1974): Approximation methods for electronic filter design. McGraw-Hill Book Company, NY.
24. Davidor, Y. (1991): A genetic algorithm applied to robot trajectory generation. Handbook of Genetic Algorithms, L. Davis (Eds), 144–165.
25. Davis, L. (1985): Job shop scheduling with genetic algorithms. Proc. 1st Int. Conf. Genetic Algorithms, J.J. Grefenstette (Eds), 136–140.
26. Davis, L. (1989): Adapting operator probabilities in genetic algorithms. Proc. 3rd Int. Conf. Genetic Algorithms, 61–69.
27. Davis, L. (1991): Handbook of genetic algorithms. Van Nostrand Reinhold.
28. Davies, R. and T. Clarke (1995): Parallel implementation of a genetic algorithm. Control Eng. Practice, **3(1)** 11–19.
29. Deb K. and D.E. Goldberg (1991): Analyzing deception in trap functions. Technical Report IlliGAL 91009, Department of Computer Science, University of Illinois at Urbana-Champaign, Urbana.
30. DeJong, K. (1975): The analysis and behaviour of a class of genetic adaptive systems. PhD thesis, University of Michigan.
31. DeJong, K.A. and W.M. Spears (1990) : An analysis of the interacting roles of population size and crossover in genetic algorithms. Proc. First Workshop Parallel Problem Solving from Nature, Springer Verlag, Berlin, 38–47.
32. Dodd, N., D. Macfarlane and C. Marland (1991): Optimization of artificial neural network structure using genetic techniques implemented on multiple transputers. Transputing '91, **2** 687–700.
33. Dyann, W.S. and R. Tjian (1985): Control of eukaryotic messenger RNA synthesis by sequence-specific DNA-binding proteins. Nature **316**, 774–778.
34. Elliott, S.J., P.A. Nelson, I.M. Stothers and C.C. Boucher (1990): In-flight experiments on the active control of propeller-induced cabin noise. J. Sound and Vibration, **140**, 219–238.
35. Elliott, S.J. and P.A. Nelson, (1993): Active noise control. IEEE Signal Processing Magazine, Oct, 12–35.
36. Eriksson, L.J. (1991): Development of the filtered-U algorithm for active noise control. J. Acoustic Soc. Am **89**, 257–265.
37. Eshelman, L.J., R. Caruna, and J.D. Schaffer (1989): Biases in the crossover landscape. Proc. 3rd Int. Conf. Genetic Algorithms, 10–19.
38. Fitzpatrick, J.M. and J.J. Grefenstette (1988): Genetic algorithms in noisy environments. Machine Learning, **3(2/3)**, 101–120.
39. Fonseca, C.M., E.M. Mendes, P.J. Fleming and S.A. Billings (1993): Non-linear model term selection with genetic algorithms. Proc. Workshop on Natural Algorithms in Signal Processing, 27/1–27/8.

40. Fonseca, C.M. and P.J. Fleming (1993): Genetic algorithms for multiobjecitve optimization: formulation, discussion and generalization. Proc. 5th Int. Conf. Genetic Algorithms, (S. Forrest, ed.), 416–423.
41. Fonseca, C.M. and P.J. Fleming (1994): An overview of evolutionary algorithms in multiobjective optimization. Research Report No. 527, Dept. of Automatic Control and Systems Eng., University of Sheffield, UK.
42. Fonseca, C.M. and P.J. Fleming (1995): Multiobjecitve genetic algorithms made easy: selection, sharing and mating restriction. Proc. 1st IEE/IEEE Int. Conf. on GAs in Engineering Systems: Innovations and Applications, 45–52.
43. Fourman, M.P. (1985): Compaction of symbolic layout using genetic algorithm. Proc. 1nd Int. Conf. Genetic Algorithms, 141–153.
44. Fu, L.M. (1994): Neural networks in computer intelligence. McGraw-Hill.
45. Garey, M.R. and D.S. Johnson (1979): Computers and intractability: a guide to the theory of NP-completeness. Freeman, San Francisco.
46. Gill, P.E., W. Murray and M.H. Wright (1981): Practical optimization. Academic Press.
47. Gillies, A.M. (1985): Machine learning procedures for generating image domain feature detectors. Doctoral Dissertation, University of Michigan.
48. Glover, K. and D. McFarlane (1989): Robust stabilization of normalized coprime factor plant descriptions with -bounded Uncertainty. IEEE Trans. Automat. Contr., **AC-34(8)** 821–830.
49. Goldberg, D.E. (1987): Simple genetic algorithms and the minimal deceptive problem. Genetic Algorithms and Stimulated Annealing, L. Davis (Ed.) 74–88.
50. Goldberg, D.E. (1989): Genetic algorithms in search, optimization and machine learning. Addison-Wesley.
51. Goldberg, D.E. (1990): Real-coded genetic algorithms, virtual alphabets, and block. Technical Report No. 90001, University of Illinois.
52. Goldberg, D.E. and R. Lingle (1985): Alleles, locis, and the traveling salesman problem. Proc. Int. Conf. Genetic Algorithms and Their Applications, 154–159.
53. Goldberg, D.E. and J.J. Richardson (1987): Genetic algorithms with sharing for multimodal function optimization. Proc. 2nd Int. Conf. Genetic Algorithms, 41–47.
54. Goldberg, D.E. and R.E. Smith (1987): Nonstationary function optimization using genetic dominance and diploidy. Proc. 2nd Int. Conf. Genetic Algorithms, 59–68.
55. Gordon, V. and D. Whitley (1993): Serial and parallel genetic algorithms as function optimizer. Proc. 5th Int Conf. Genetic Algorithms, 177–183.
56. Gorges-Schleuter, M. (1989): ASPARAGOS An asynchronous parallel genetic optimization strategy. Proc. 3rd Int. Conf. Genetic Algorithms, 422–427.
57. Grefenstette, J.J. (1986): Optimization of control parameters for genetic algorithms. IEEE Trans Systems, Man, and Cybernetics, **SMC-16(1)**, 122–128.
58. Grefenstette J.J. and J. Baker (1989): How genetic algorithms work: A critical look at implicit parallelism. Proc 3rd Int. Conf. Genetic Algorithm.
59. Grefenstette J.J. (1990): A user's guide to GENESIS v5.0. Naval Research Laboratory, Washington, D.C.
60. Grefenstette, J.J. (1992): Genetic algorithms for changing environments. Parallel Problem Solving from Nature, 2, 137–144.
61. Grefenstette, J.J. (1993): Deception considered harmful. Foundations of Algorithms, 2, L. Darrell Whitley (Ed.) 75–91.

62. Guillemin, E.A. (1956): Synthesis of passive networks. John Wiley and Sons, NY.
63. Hajela, P. and Lin, C.-Y. (1992): Genetic search strategies in multicriterion optimal design. Structural Optimization **4** 99–107.
64. Hall, H.R., W.B. Ferren and R.J. Bernhard (1992): Active control of radiated Sound from ducts. Trans. of the ASME, **114** 338–346.
65. Heitkoetter, J. and D. Beasley (Eds) (1994): The Hitch-Hiker's guide to evolutionary computation: A list of frequently asked questions (FAQ). USENET:comp.ai.genetic., 1994.
66. Ho, K.C., Y.T. Chan and P.C. Ching (1993): Adaptive time-delay estimation in nonstationary signal and/or noise power environments. IEEE Trans. Signal Processing, **41(7)**, 2289–2299.
67. Ho, Y.C., K.F. Man, K.P. Cheuk and K.T. Ng (1994): A fully automated water supply system for high rise building. Proc. 1st Asian Control Conference, 1–4.
68. Ho, Y.C., K.F. Man, K.S. Tang and C.Y. Chan (1996): A dependable parallel architecture for active noise control. IFAC World Congress 96 (to be published).
69. Hollstien, R.B. (1971): Artificial genetic adaptation in computer control systems. PhD thesis, University of Michigan.
70. Holland, J.H. (1975): Adaption in natural and artificial systems. MIT Press.
71. Homaifar, A. and Ed McCormick (1995): Simultaneous design of membership functions and rule sets for fuzzy controllers using genetic algorithms. IEEE Trans Fuzzy Systems **3(2)**, 129–139.
72. Horn, J. and N. Nafpliotis (1993): Multiobjective optimization using the niched pareto genetic algorithm. IlliGAL Report 93005, University of Illinois at Urbana-Champaign, Urbana, Illinois, USA.
73. Hoyle, D.J., R.A. Hyde and D.J.N. Limebeer (1991): An approach to two degree of freedom design. Proc. 30th IEEE Conf. Dec. Contr., 1581–1585.
74. Itakura, F. (1975): Minimum prediction residual principle applied to speech recognition. IEEE Trans. Acoust., Speech, Signal Processing **ASSP-23**, 67–72.
75. Jakob, W., M. Gorges-Schleuter and C. Blume (1992): Application of genetic algorithms to task planning and learning. Parallel Problem Solving from Nature, 2, 291–300.
76. Jang, J.-S.R. and C.-T. Sun (1995): Neuro-fuzzy modeling and control. Proc. IEEE, **83(3)**, 378–406.
77. Janikow, C.Z. and Z. Michalewicz (1991): An experimental comparison of binary and floating point representations in genetic algorithms. Proc. 4th Int. Conf. Genetic Algorithms, 31–36.
78. Jones, K.A., J.T. Kadonga, D.J. Rosenfeld, T.J. Kelly and R. Tjian (1987): A cellular DNA binding protein that activates eukaryotic transcription and DNA replication. Cell **48**, 79–84.
79. Karr, C.L. (1991): Genetic algorithms for fuzzy controllers. AI Expert, **6(2)**, 26–33.
80. Karr, C.L. and E.J. Gentry (1993): Fuzzy control of pH using genetic algorithms. IEEE Trans Fuzzy Systems **1(1)**, 46–53.
81. Kennedy, S.A. (1991): Five ways to a smarter genetic algorithm. AI Expert, Dec, 35–38.
82. Kido, K., M. Abe and H. Kanai (1989): A new arrangement of additional sound source in an active noise control system. Proc. Inter-Noise 89, 483–488.
83. Kornberg, A. (1980): DNA replication. Freeman, San Francisco.

84. Kozek, T., T. Roska and L. Chua (1993): Genetic algorithm for CNN template learning. IEEE Trans. on Circuit and Systems - I : Fundamental Theory and Applications **40(6)**.
85. Kröger, B., P. Schwenderling and O. Vornberger (1993): Parallel genetic packing on transputers. Parallel Genetic Algorithms: Theory and Applications, Amsterdam: IOS Press, 151–185.
86. Kwong, S., Q. He and K.F. Man: Genetic time warping for isolated word recognition. International Journal of Pattern Recognition and Artificial Intelligence (to be published).
87. Lam, H.Y.-F. (1979): Analog and digital filters: design and realization. Prentice-Hall, Englewood Cliffs, NJ.
88. Limebeer, D.J.N. (1991): The Specification and purpose of a controller design case study. Proc. 30th IEEE Conf. Dec. Contr., Brighton, England, 1579–1580.
89. Leug, P. (1936): Process of silencing sound oscillations. U.S. Patent No. 2,043,416.
90. Louis, S.J. and Rawlins, G.J.E. (1993): Pareto optimality, GA-easiness and deception. Proc. 5th Int. Conf. Genetic Algorithms, 118–223.
91. Mahfoud, S.W. (1992): Crowding and preselection revisited. IlliGAL Report No. 92004, Department of Computer Science, Univeristy of Illinois at Urbana-Champaign.
92. Mahfoud, S.W. (1994): Population sizing for sharing methods. IlliGAL Report No. 94005, Department of Computer Science, University of Illinois at Urbana-Champaign, Urbana.
93. Man, K.F., K.S. Tang and S. Kwong (1996): Genetic algorithms: concept and applications. IEEE Trans. Industrial Electronics (to be published).
94. Manderick, B. and P. Spiessens (1989): Fine-grained parallel genetic algorithms. Proc. 3rd Int. Conf. Genetic Algorithms, 428–433.
95. Mangasarian, O. L. and W. H. Wolberg (1990): Cancer diagnosis via linear programming. SIAM News **23(5)**, 1–18.
96. Maniatis, T., S. Goodbourn, J.A. Fischer (1987): Regulation of inducible and tissue-specific gene expression. Science **236**, 1237–1245.
97. Maniezzo, V. (1994): Genetic evolution of the topology and weight distribution of neural networks. IEEE Trans. Neural Networks **5(1)**, 39–53.
98. MATHWORKS (1991): MATLAB user's guide. The MathWorks, Inc.
99. McFarlane, D.C. and K. Glover (1990): Robust controller design using normalized coprime factor plant descriptions. Lecture Notes Control & Information Sciences, **138**, Berlin:Springer-Verlag.
100. McFarlane, D.C. and K. Glover (1992): A Loop Shaping design procedure using synthesis. IEEE Trans. Auto. Control, **AC-37(6)** 749–769.
101. Michalewicz, Z. (1994): Genetic Algorithms + Data Structures = Evolution Program. 2nd Ed., Springer-Verlag.
102. Miller, G.F., P.M. Todd, and S.U. Hegde (1989): Designing neural networks using genetic algorithms. Proc. 3rd Int. Conf. Genetic Algorithms, 379–384.
103. Montana, D.J. and L. Davis (1989): Training feedforward neural networks using genetic algorithms. Proc. 11th Joint Conf. on Artificial Intelligence, **IJCAI-11**, 762–767.
104. Mühlenbein, H. (1989): Parallel genetic algorithms, population genetics and combinatorial optimization. Parallelism, Learning, Evolution, Springer-Verlag, 398-406.
105. Munetome, M., Y. Takai and Y. Sato (1993): An efficient migration scheme for subpopulation-based asynchronously parallel genetic algorithms. Proc. 5th Int. Conf. Genetic Algorithms, 649.

106. Munakata, T. and Yashvant Jani (1994): Fuzzy systems: An overview. Communications of the ACM, **37(3)**, 69–76.
107. Nambiar, R. and P. Mars, (1993): Adaptive IIR filtering using natural algorithms. Proc. Workshop on Natural Algorithms in Signal Processing, 20/1–20/10.
108. Omlin, C.W., and C.L. Giles (1993): Pruning recurrent neural networks for improved generalization performance. Tech. Report No. 93-6, Computer Science Department, Rensselaer Polytechnic Institute.
109. Palmer, C.C. and A. Kershenbaum (1995): An approach to a problem in network design using genetic algorithms. Networks **26**, 151–163.
110. Park, D., A. Kandel and G. Langholz (1994): Genetic-based new fuzzy reasoning models with application to fuzzy control. IEEE Trans Systems, Man and Cybernetics **24(1)**, 39–47.
111. Park, Y., and H. Kim (1993): Delayed-X algorithm for a long duct system. Proc. Inter-Noise 93, 767–770.
112. Parlos, A.G., B. Fernandez, A.F. Atiya, J. Muthusami and W.K. Tsai (1994): An accelerated learning algorithm for multilayer perceptron networks. IEEE Trans. Neural Networks **5(3)**, 493–497.
113. Procyk, T.J. and E.H. Mamdani (1979): A linguistic self-orgainizing process controller. Automatica **15**, 15–30.
114. Ptashne M. (1986): Gene regulation by proteins acting nearby and at a distance. Nature **322**, 697–701.
115. Rabiner, L.R. and B.H. Juang (1994): Fundamentals of speech recognition. Prentice-Hall.
116. Radding, C. (1982): Homologous pairing and strand exchange in genetic recombination. Annual Review of Genetics **16**, 405–437.
117. I. Rask and C.S. Downes (1995): Genes in medicine. Chapman & Hall.
118. Reed, F.A., P.L. Feintuch, and N.J. Bershad (1981): Time-delay estimation using the LMS adaptive filter-static behavior. IEEE Trans. Acoust., Speech, Signal Processing **ASSP-29**, 561–568.
119. Richardson, J.T., M.R. Palmer, G. Liepins and M. Hilliard (1989): Some guidelines for genetic algorithms with penalty functions. Proc. 3rd Int. Conf. Genetic Algorithms, 191–197.
120. Rosenberg, A.E. (1976): Evaluation of an automatic speaker verification system over telephone line. Bell Syst. J. **55**, 723–744.
121. Rudell, R. and R. Segal (1989): Logic synthesis can help in exploring design choice. 1989 Semicustom Design Guide, CMP Publications, Manhasset, NY.
122. Rumelhart, D.E., G.E. Hinton and R.J. Williams (1986): Learning internal representations by error propagation. Parallel Distributed Processing: Explorations in the Microstructures of Cognition, D.E. Rumelhart and J.L. McLelland, Eds. Cambridge, MA: MIT Press, 318–362.
123. Saenger, W. (1984): Principles of nucleic acid structure. Springer Verlag, New York.
124. Sakoe, H. and S. Chiba (1970): A similarity evaluation of speech patterns by dynamic programming. presented at the Dig. 1970 Nat. Meeting, Inst. Electron. Comm. Eng. Japan, 136.
125. Sakoe, H. and S. Chiba (1978): Dynamic programming algorithm optimization for spoken word recognition. IEEE Trans. Acoust., Speech, Signal Processing **ASSP-26**, 43–49.
126. Schaffer, J.D. (1985): Multiple objective optimization with vector evaluated genetic algorithms. Proc. 1st Int. Conf. Genetic Algorithm, 93–100.
127. Serfling E., M. Jasmin, and W. Schaffner (1985): Enhancers and eukaryotic gene transcription. Trends in Genetics **1** 224–230.

128. Sharpe, R.N., M.Y. Chow, S. Briggs and L. Windingland (1994): A Methodology using fuzzy logic to optimize feedforward artificial neural network configurations. IEEE Trans. Systems, Man and Cybernetics **24(5)**, 760–768.
129. Shynk, J.J. (1989): Adaptive IIR filtering. IEEE ASSP Magazine **April**, 4–21.
130. Simpson, P.K. (1990): Artificial neural systems: Foundations, paradigms, applications, and implementations. Pergamon Press, 100–135.
131. Simth D. (1985): Bin packing with adaptive search. Proc. Int. Conf. Genetic Algorithms and Their Applications, 202–206.
132. Skogestad, S., M. Morari and J.C. Doyle (1988): Robust control of ill-conditioned plants: High-purity distillation. IEEE Tran. Auto. Control **AC-33(12)**, 1092–1105.
133. So, H.C., P.C. Ching, and Y.T. Chan (1994): A new algorithm for explicit adaptation of time delay. IEEE Trans Signal Processing **42(7)**, 1816–1820.
134. Spears, W.M. and K. DeJong (1991): An analysis of Multi-point crossover. Foundations of Genetic Algorithms, G.J.E. Rawlins (Eds), 301–315.
135. Srinivas, M. and L. M. Patnaik (1994): Genetic algorithms: a survey. Computer, June, 17–26.
136. Sutton, T.J., S.J. Elliott and A.M. McDonald (1994): Active control of road noise insider vehicles. Noise Control Eng. J. **42 (4)**, 137–147.
137. Syswerda, G. (1989): Uniform crossover in genetic algorithms. Proc. 3rd Int. Conf. Genetic Algorithms, 2–9.
138. Syswerda, G. (1991): Schedule optimization using genetic algorithms. Handbook of Genetic Algorithms, 332–349.
139. Szostak, J., T.L. Orr-Weaver, R.J. Rothstein, F.W. Stahl (1983): The double-strand-break repair model for recombination. Cell **33**, 25–35.
140. Naofumi Takagi, Hiroto Yasuura and Shuzo Yajima, (1985): High-speed VLSI multiplication algorithm with a redundant binary addition tree. IEEE Trans Computers **C-34(9)**, 789–796.
141. Tamaki, H. and Y. Nichikawa (1992): A paralleled genetic algorithm Based on a neighborhood model and its application to job shop scheduling. Parallel Problem Solving from Nature, 2, 573–582.
142. Tang, K.S., K.F. Man and C.Y. Chan (1994): Fuzzy control of water pressure using genetic algorithm. Proc IFAC Workshop on Safety, Reliability and Applications of Emerging Intelligent Control Technologies, 15–20.
143. Tang, K.S., K.F. Man and S. Kwong (1995): GA approach to time-variant delay estimation. Proc. Int. Conf. on Control and Information, 173–175.
144. Tang, K.S., K.F. Man, C.Y. Chan, S. Kwong, P.J. Fleming (1995): GA approach to multiple objective optimization for active noise control. IFAC Algorithms and Architectures for Real-Time Control AARTC 95, 13–19.
145. Tang, K.S., C.Y. Chan, K.F. Man and S. Kwong (1995): Genetic structure for NN topology and weights optimization. 1st IEE/IEEE Int. Conf. on GAs in Engineering Systems: Innovations and Applications, Sheffield, UK, 250–255.
146. Tang, K.S., K.F. Man, S. Kwong, C.Y. Chan and C.Y. Chu (1996): Application of the genetic algorithm to real-time active noise control. Journal of Real-Time Systems (to be published).
147. Tang, K.S., C.Y. Chan, K.F. Man (1996): Hierarchical genetic algorithm for fuzzy controller design. IEEE Int. Conf. on Industrial Technology, Shanghai China, (to be published).
148. Tang, K.S., K.F. Man, S. Kwong and Q. He (1996): Genetic algorithms and their applications in signal processing. IEEE Signal Processing Magazine (to be published).

149. Tang, Y.C. (1994): Tolkien reference manual. Dept. of Computer Science, The Chinese University of Hong Kong.
150. Tanse, R. (1989): Distributed genetic algorithms. Proc. 3rd. Int. Conf. Genetic Algorithms, 434–439.
151. Thomas, B. (1992): Users guide for GENEsYs. System Analysis Research Group, Dept. of Computer Science, University of Dortmund.
152. Velichko, V.M. and N.G. Zagoruko (1970): Automated recognition of 200 words. Int. J. Man-Machine Stud. **vol. 2**, 223.
153. Weinberg, L. (1975): Network analysis and synthesis. R.E. Kreiger, Huntington, NY.
154. Whidborne, J.F., I.Postlethwaite and D.W. Gu (1994): Robust controller design using loop-shaping and the method of inequalities. IEEE Trans Control System Technology. **2**(4) 455–461.
155. Whidborne, J.F., D.W. Gu and I. Postlethwaite (1995): Algorithms for solving the method of inequalities - a comparative study. Proc. American Control Conference, (to be published).
156. White, G. and R. Neely (1976): Speech recognition experiments with linear prediction, bandpass filtering, and dynamic programming. IEEE Trans. Acoust., Speech, Signal Processing **ASSP-24**, 183–188.
157. White, M.S. and S.J. Flockton (1993): A comparative study of natural algorithms for adaptive IIR filtering. Workshop on Natural Algorithms in Signal Processing, 22/1–22/8.
158. Whitley, D. (1989): The GENITOR algorithm and selection pressure: Why rank-based allocation of reproductive trials is best. Proc. 3rd Int. Conf. Genetic Algorithms (J.D. Schaffer, Ed.) 116–121.
159. Whitley, D. (1993): A genetic algorithm tutorial. Technical Report CS-93-103, Department of Computer Science, Colorado State University.
160. Widrow, B. and S.D. Stearns (1984): Adaptive signal processing. Prentice Hall.
161. Widrow, B., D.E. Rumelhart and M.A. Lehr (1994): Neural networks: applications in industry, business and science. Communication of the ACM **37(3)**, 93–105.
162. Wienke, D., C. Lucasius and G. Kateman (1992): Multicriteria target vector optimization of analytical procedures using a genetic algorithm. Part I. Theory, numerical simulations and application to atomic emission spectroscopy. Analytica Chimica Acta, **265(2)**, 211–225.
163. Wolberg, W.H., and O.L. Mangasarian (1990): Multisurface method of pattern separation for medical diagnosis applied to breast cytology. Proc. of the National Academy of Sciences, **87**, 9193–9196.
164. Wright, A.H. (1991): Genetic algorithms for real parameter optimization. Foundations of Genetic Algorithms, J.E. Rawlins (Ed.), Morgan Kaufmann, 205–218.
165. Youm, D.H., N. Ahmed, and G.C. Carter. (1982)On using the LMS algorithm for delay estimation. IEEE Trans. Acoust., Speech, Signal Processing, **ASSP-30**, 798–801.
166. Yuan, Z.D. and X. Wang (1994): Determining the node number of neural network models. IFAC Workshop on Safety, Reliability and Applications of Emerging Intelligent Control Technologies, 60–64.
167. Zadeh, L.A. (1973): Outline of a new approach to the analysis complex systems and decision processes. IEEE Trans. Syst., Man, Cybernetics **SMC3**, 28–44.
168. Zakian, V. and U. Al-Naib (1973): Design of dynamical and control systems by the method of inequalities. Proc. IEE, **120(11)** 1421–1427.

169. Zhang, J. (1992): Selecting typical instances in instance-based learning. Proc. of the 9th International Machine Learning Conference, 470–479.

INDEX